D0930385

Six Sigma—
The First 90 Days

Six Sigma—

THE FIRST 90 DAYS

Stephen A. Zinkgraf

With contributions from Sigma Breakthrough Technologies, Inc., Instantis, Inc., and Peak Performance Consulting Corporation

PRENTICE HALL

Upper Saddle River, NJ • Boston • Indianapolis • San Francisco
New York • Toronto • Montreal • London • Munich • Paris • Madrid
Cape Town • Sydney • Tokyo • Singapore • Mexico City

The publisher offers excellent discounts on this book when ordered in quantity for bulk purchases or special sales, which may include electronic versions and/or custom covers and content particular to your business, training goals, marketing focus, and branding interests. For more information, please contact:

U.S. Corporate and Government Sales
(800) 382-3419
corpsales@pearsontechgroup.com

For sales outside the United States, please contact:

International Sales
international@pearsoned.com

This Book Is Safari Enabled

The Safari® Enabled icon on the cover of your favorite technology book means the book is available through Safari Bookshelf. When you buy this book, you get free access to the online edition for 45 days. Safari Bookshelf is an electronic reference library that lets you easily search thousands of technical books, find code samples, download chapters, and access technical information whenever and wherever you need it.

To gain 45-day Safari Enabled access to this book:

Go to http://www.awprofessional.com/safarienabled

Complete the brief registration form

Enter the coupon code 42NE-CTTL-BRFX-KHLA-1H7Y

If you have difficulty registering on Safari Bookshelf or accessing the online edition, please e-mail customer-service@safaribooksonline.com.

Visit us on the Web: www.prenhallprofessional.com

Pearson Education, Inc.
Rights and Contracts Department
One Lake Street
Upper Saddle River, NJ 07458
Fax: (201) 236-3290

ISBN 0-13-168740-9
Text printed in the United States on recycled paper at R.R. Donnelley & Sons in Crawfordsville, IN.
First printing, April 2006

Library of Congress Cataloging-in-Publication Data

Zinkgraf, Stephen A. (Stephen Arno), 1948-
Six sigma : the first ninety days / Stephen A. Zinkgraf ; contributors, Daniel M. Kutz ... [et al.].
p. cm.
ISBN 0-13-168740-9 (hardback : alk. paper) 1. Six sigma (Quality control standard) I. Kutz, Daniel M. II. Title.
HD62.15.Z57 2006
658.4'013—dc22

2005036553

This book is dedicated to my wife, Patricia, and my son and daughter, Bill and Cindy. Their support through my hard times was directly responsible for this book. I also dedicate this book to the staff and consultants of Sigma Breakthrough Technologies, Inc., for innovative work and ideas.

Thank you all and God bless.

Contents

Preface xiii

Acknowledgments xvii

About the Author xix

Contributors List xxi

PART I PRE-LAUNCH 1

Chapter 1 Six Sigma—The Initiative, the Deployment, the Technology 3

Many Dimensions of Six Sigma 5
Six Sigma—The Initiative 6
Six Sigma—The Alignment 6
Six Sigma—The Discipline 7
Six Sigma—Leadership Development 8
Six Sigma—The Methodology 10
  Selecting the Right Projects 10
  Roles and Responsibilities of the Primary Players—Belts and Champions 10
  The Right Roadmap and Tools 12
  The Right Results 12
The Playbook 12
  Section 1: Pre-Launch Preparation 13
  Section 2: The Launch: First 90 Days 13
  Section 3: Post-Launch Actions 15

Chapter 2 The True Nature of Six Sigma: The Business Model 17
The Business Model 17
Assessing External Realities 18
Setting Financial Targets 19
The Nature of Six Sigma 19
Linking Internal Activities 22
Focusing On, Finding, and Delivering the Money 24
Focusing on the Money 24
Finding the Money 24
Delivering the Money 24
Focusing on the Money 25
Six Sigma Evolution 25
Six Sigma: Focus on Money 27
Finding the Money 28
Analyzing Profits 32
Delivering the Money 35

Chapter 3 Six Sigma Launch Philosophy 37
The Nature of a Six Sigma Deployment 38
Six Sigma Deployment Timing 40
Kotter's Philosophical Deployment Architecture 41
Kotter Stage 1: Establishing a Sense of Urgency 42
Kotter Stage 2: Creating a Guiding Coalition 45
Kotter Stage 3: Creating a Vision and Strategy 47
Kotter Stage 4: Communicating the Change Vision 48
Kotter Stage 5: Empowering Broad-Based Change 50
Kotter Stage 6: Generating Short-Term Wins 51
Kotter Stage 7: Consolidating Gains and Producing More Change 51
Kotter Stage 8: Anchoring New Approaches in the Culture 52
Summary 53

Chapter 4 Getting Early Support: Selecting a Six Sigma Provider 55
Six Sigma, Consultants, and Consulting Costs 57
Creating a Partnership 59
Identifying, Prioritizing, and Selecting Providers 60
Differentiating Providers 63
Questions to Ask Providers 66
Terms of Engagement 67
Summary 68

Chapter 5 Strategy: The Alignment of External Realities, Setting Measurable Goals, and Internal Actions 69

Chapter 6 Defining the Six Sigma Program Expectations and Metrics 87

Defining the Bottom-Line Six Sigma Program Expectations and Metrics 90

Bottom-Line Expectations—Productivity and Efficiency Improvement 91

Top-Line Expectations—Effectiveness and Value Proposition 99

Six Sigma Participation Expectations 100

PART II THE FIRST 90 DAYS 103

Chapter 7 Defining the Six Sigma Project Scope 105

Connecting Six Sigma to Other Initiatives 106

Pilot or Full Deployment? 108

By Business Unit or Division? 109

Deployment Domestically or Globally? 110

Six Sigma Program Content 111

Scheduling Events 113

Chapter 8 Defining the Six Sigma Infrastructure 117

The Six Sigma Infrastructure 119

CEO and Executive Team 120

Six Sigma Initiative Champion 123

Six Sigma Deployment Champions—Business Unit 126

Six Sigma Project Champions 128

Master Black Belts 130

Six Sigma Black Belts 133

Six Sigma Green Belts 135

Six Sigma Project Team Members 138

Six Sigma Finance Support 139

Six Sigma Human Resource Support 141

Six Sigma Project Tracking 142

Six Sigma Steering Teams 142

Six Sigma Certifications 144

Six Sigma Infrastructure Summary 146

Chapter 9 Committing to Project Selection, Prioritization, and Chartering 149

Six Sigma Leadership Roadmap 150

Selecting the Right Projects 151

	Project Prioritization and Selection Roadmap	157
	Types of Projects: Top-Down and Bottom-Up	161
	Prioritizing Projects	165
	Checking and Establishing Accountability: Chartering Projects	176
	Summary of Six Sigma Project Selection	182
Chapter 10	**Creating Six Sigma Executive and Leadership Workshops**	**185**
	Executive Team Six Sigma Workshop	186
	A Generic Outline for the Executive Team Workshop	188
	Actual Three-Day Agenda	189
	Business Team Six Sigma Workshop	190
	A Generic Outline for the Business Team Workshop	191
	One-Day Purpose and Agenda—Sample of an Actual Workshop	192
	Actual Four-Day Business Team Agenda	193
	Deployment and Project Champion Workshops	195
	A Generic Outline for the Deployment and Project Champion Workshops	196
	Actual Example of the Deployment and Project Champion Workshops	197
	A Process for Rapidly Customizing Executive/Champion Workshops	199
	Q1. What are the three highest-priority issues facing your business today?	202
	Q2. What is your personal experience in deploying Lean Six Sigma?	202
	Q3. What is your personal knowledge of Six Sigma methodology and tools?	203
	Q4. What would you personally like to get out of this session?	203
	Materials Review with Key Stakeholders	204
	Being Successful, Avoiding Problems	205
Chapter 11	**Selecting and Training the Right People**	**209**
	Six Sigma, Lean Sigma, and Lean Training Portfolio—Enterprise Wide	211
	Training Portfolios: Six Sigma, Lean Sigma, and Lean Training	211
	Six Sigma and Lean	213
	When Six Sigma?	213
	When Lean?	214
	Leadership Workshops	214
	Value Creation and Management: Design for Six Sigma	215
	Chemical Design for Six Sigma—Black Belt Training	216
	Operational Black Belt—Manufacturing Process Improvement	218
	K-Sigma Black Belt (Six Sigma and Lean)—Manufacturing Process Improvement	220
	Six Sigma "Belt" Training Summary	225
	Internalization of Training	226
	The Master Black Belt	226
	Master Black Belt Training Overview	227

Training Plans 227
Black Belt Course Evaluation—Week 4 231

Chapter 12 **Communicating the Six Sigma Program Expectations and Metrics** **235**
A Communications Model 236
Creating Special Communication Systems 238
The Message 240
Six Sigma Program Deployment 244
The Media 244
Communication Plans 246

PART III **POST-LAUNCH** **253**

Chapter 13 **Creating the Human Resources Alignment** **255**
Six Sigma Organizational Structure 257
Master Black Belts 258
Black Belts 258
Green Belts 259
Talent Selections 259
Reasonable Numbers of Trainees for the Annual Forecast 263
Workshop for Forecasting Volume of Students to Be Trained 263
Workshop for Developing the Belt Selection Process 266
Product Development Black Belts: Some Criteria and Information 266
Product Development Black Belts 267
Position Profile 268
Position Title: Six Sigma Green Belt Salary Grade: TBD 268
Recognition and Rewards 274
Suggestions for Recognition Actions 275
Actual Examples of Reward and Recognition 277
Belt Retention and Career Planning 279
Workshop Addressing Retention of Belts 281
Six Sigma Deployment HR Support 281
Specialty Chemicals Vision 282

Chapter 14 **Defining the Software Infrastructure: Tracking the Program and Projects** **285**
Enterprise Management Solution Fundamentals 287
Enterprise Solutions: Getting Started 289
Enterprise Solution Requirements 290

Enterprise Solution Packages (Alphabetically Listed) 297
Desktop Solutions 298
Planning Your Solutions 300
Executing Solutions 300
Tuning and Expansion 301

Chapter 15 Leading Six Sigma for the Long Term 303
Kotter's Stages Seven and Eight 303
Five Steps to Leading Six Sigma 306
Six Sigma Leadership Step 1—Select the Right Projects 307
Six Sigma Leadership Step 2—Select and Train the Right People 312
Six Sigma Leadership Step 3—Develop and Implement Improvement Plans for Key Six Sigma Projects 318
Six Sigma Leadership Step 4—Manage Six Sigma for Excellence 319
Six Sigma Leadership Step 5—Sustain the Performance Gains 323
Six Sigma Handbooks and Other Anchors 325

Chapter 16 Reinvigorating Your Six Sigma Program 329
Critical Analysis of Your Six Sigma Deployment 331
Six Sigma Survey 332
Back to the Basics: Critical Quality Dimensions 333
Leading Six Sigma 334
Failure Modes for Selecting the Right Projects 335
Failure Modes for Selecting and Training the Right People 336
Failure Modes for Developing and Implementing Improvement Plans 337
Failure Modes for Managing for Excellence 338
Failure Modes for Sustaining the Gains 339
Is Your Deployment Effective and Efficient? 340
Is Everything Aligned to the Strategy? 341
Time to Extend Your Program: Using Deployment Successes to Go Further and Faster 346
Assessing Your Program Using Kotter's Eight Stages 352
Six Sigma: The Initiative 354

Appendix A RFP Sample Format 357

Appendix B About the Contributors 367

Index 371

Preface

The new Six Sigma has had no less than a dazzling debut, starting in late 1994 with AlliedSignal. Originated in 1987 at Motorola, Six Sigma was adopted by very few companies, though it was not taken very seriously. Only after AlliedSignal (now known as Honeywell) demonstrated Six Sigma's effectiveness in redefining a company, achieving dramatic results and—more importantly—positively affecting the lives of thousands of employees, did Six Sigma catch the imagination of corporate America. General Electric (GE) adopted Six Sigma in 1996 with a vengeance.

With Jack Welch (GE) and Larry Bossidy (AlliedSignal) promoting the effectiveness and necessity of Six Sigma, the Six Sigma tsunami began. A former Motorola quality leader, Richard Schroeder, was the chief advisor to both these dynamic leaders. In fact, Richard lead the AlliedSignal Six Sigma deployment for Larry. With the leadership of these three spiritual leaders, hundreds of companies have met the challenge of deploying Six Sigma into their businesses. There are very few stories of failed deployments.

This book was written to inspire leaders to commit to deploying Six Sigma and to demystify the process. If you are a senior leader of an organization, or a leader who might be involved in launching Six Sigma, this book is for you. The book will also work well to provide a handbook (or playbook) to support your organization's effort as you start Six Sigma. Everyone wants to know what the next great initiative will be after Six Sigma.

This book will not directly answer that question. But, the next best thing may reside in a company's or organization's ability to effectively deploy change initiatives. To quickly identify and initiate a change program will be the core competency of the millennium. This book demonstrates the way to deploy a complex change initiative within 90 days—

the deployment milestones are here, and the roadmaps are here. Essentially, if you don't effectively deploy Six Sigma, it won't matter what the next big program will be.

These milestones and roadmaps are based on my 18+ years' experience in Six Sigma, plus the extensive experience of the 10 contributors to the book. We have seen the critical milestones that work and the failure modes that lead to mediocrity. More importantly, we have provided enough detail to equip you with the knowledge of what to do, how to do it, and when to do it.

Because of my Six Sigma work with the 1994 AlliedSignal deployment, I draw many principles from Larry Bossidy and Richard Schroeder. Larry was the business leader and Rich was the spiritual leader for the initiative—they were a perfect team. These two leaders were responsible for, most likely, the best deployment of Six Sigma ever. With pinpoint accuracy, their leadership teams deployed a very large program in about 90 days. This deployment was a tribute to leaders who weren't afraid to get emotionally involved with change.

Six Sigma presents a clear challenge because of its multifaceted nature. The Six Sigma continuum includes sophisticated statistical methods, problem-solving roadmaps, alignment of strategy to internal actions, and high-powered leadership. This book addresses the actions necessary to create hundreds of "Great Groups" working throughout your company to redefine your performance.

This book will consist of three sections: Part I, "Pre-Launch," Part II, "The First 90 Days," and Part III, "Post-Launch." The term *launch* is defined as the activities leading up to the first wave of Six Sigma training programs.

PART I: PRE-LAUNCH

The pre-launch preparation chapters will provide you with the specific actions and milestones necessary to get to the starting line.

Chapter 1 provides you with an executive overview of Six Sigma as an initiative. Because Six Sigma is complex, I spend little time on the problem-solving methodology and more time on deployment milestones. **Chapter 2** provides you with insights into the process of focusing on, finding, and delivering the money or results. I also address the business model to ensure alignment between your Six Sigma activities and your balance sheet. **Chapter 3** gives you an executive overview summarizing the nature of a typical Six Sigma deployment. My team and I have sifted through some 50 Six Sigma launches to find the best practices.

Chapter 4 supplies you with the rationale for using an external consulting group to support your deployment and a list of requisites to consider when evaluating providers. I also provide a quantitative decision matrix and a draft of a request for proposal.

Chapter 5 delineates the process by which projects are linked to the organizational strategy. This chapter describes the process by which projects are selected and prioritized to obtain a direct line of site to the strategy.

Chapter 6 discusses how to measure the success of your Six Sigma launch with metrics that are directly tied to your strategic and annual operating plans. This chapter instructs you in the way to develop your strategic metrics and, more importantly, how to develop aggressive but achievable goals.

PART II: THE FIRST 90 DAYS

This section covers the time from the initial executive training workshop to the launch of the first wave (class) of Black Belt or Green Belt training. This section provides specific actions and milestones that lead to a successful Six Sigma launch. Each chapter provides a step-by-step roadmap to accomplish a specific set of actions and provides recommended timing. Launching a Six Sigma program in 90 days is realistic. Very large companies such as Honeywell and 3M have done so, along with several smaller companies. In fact, 3M beat the 90-day timeframe by about 60 days.

Chapter 7 helps you decide what your Six Sigma deployment will look like. The chapter recommends the way you can tie Six Sigma to earlier initiatives, and compares the advantages and disadvantages of pilot projects versus full-scale organization-wide deployments. I discuss approaches that consider division-to-division and geographic deployments. In addition, the chapter reviews the multitude of Six Sigma and Lean programs that are available to you to deploy in your organization. Depending on your strategic requirements, this chapter advises you on which programs to launch and in what order.

Chapter 8 outlines the roles and responsibilities of the key players in a Six Sigma launch. It also offers selection criteria and timing. **Chapter 9** recommends ways the very critical leadership workshops should be developed and formatted. This chapter presents several different approaches to each workshop, along with sample agendas. **Chapter 10** is one of the most important chapters in the book. This chapter provides a step-by-step methodology by which breakthrough projects are identified, prioritized, and chartered.

Chapter 11 reviews the roles of the significant Six Sigma players and details the training required to fulfill these roles. The chapter also discusses the pros and cons of hiring external personnel and developing internal personnel. The chapter also provides insights into the process of internalizing the Six Sigma training. **Chapter 12** provides ideas for developing a comprehensive communications plan. Emphasizing the who, what, when, where, why, how, and how much, this chapter discusses the coordination of various communication channels. This chapter provides you with the essence of marketing Six Sigma throughout your organization.

PART III: POST-LAUNCH

This section discusses the means and methods for institutionalizing your Six Sigma program for the long term. We provide a leadership roadmap along with specific actions that must occur every year for the program to last. Milestones such as compensation plans, career ladders, and software support are detailed.

Chapter 13 provides guidance in aligning Human Resource systems to the Six Sigma program. This chapter discusses organization structure, succession planning, career planning, measuring Belt performance, and reward and recognition. **Chapter 14** provides an overview of the requirements of a project-tracking software system. The chapter also provides some examples of systems I have encountered over the years. **Chapter 15** provides a detailed leadership roadmap that, if followed, ensures the long-term success of your Six Sigma program. This step-by-step approach is based on our work with over 40 companies over the last 15 years.

Chapter 16 provides methods for assessing your current deployment and reinvigorating it based on identified gaps in the deployment. Success in deploying Six Sigma provides the opportunity to launch successive change initiatives.

This book is designed to provide you with a turnkey method to launch a Six Sigma initiative in 90 days. By using this fast approach, you will be able to quickly and effectively align your people to your strategy, processes, and customers. By doing it within 90 days, you ensure that you will have a different company within 12 months. These fast launches also demonstrate the requirement of the full commitment of your leadership team. These 90 days will be the most memorable of your career.

Acknowledgments

First, I acknowledge Richard Schroeder for paving the way for all of us. His courage, conviction, and leadership were the catalysts for the current Six Sigma movement. I am pleased to acknowledge all the employees of Sigma Breakthrough Technologies, Inc. (SBTI). Without them, this project would not have happened. I also want to acknowledge the efforts of SBTI's consultants for the great work they've done for the past eight years.

About the Author

Dr. Stephen Zinkgraf, CEO of Sigma Breakthrough Technologies, Inc., holds a Ph.D. in Educational Psychology from Texas A&M University. After spending a number of years in academia, Steve joined Motorola in 1988 as a statistical quality engineering group leader, leading a large electronics plant's SPC effort. Other positions held while at Motorola include engineering group leader, production manager, and plant quality manager. He has also spent time at Compaq Computer Company as a staff statistician and at Asea, Brown, Bovari, Ltd. (ABB), leading an internal consulting group focused on implementing Six Sigma in combination with Lean Manufacturing systems. Steve spent three years at AlliedSignal, leading the deployment of Operational Excellence (Six Sigma) in AlliedSignal's $4 billion Engineered Materials Sector, yielding over $350 million in pretax income in three years. In May 1997, Steve founded SBTI to drive business process excellence around the world.

Contributors List

Sigma Breakthrough Technologies, Inc. (SBTI)

Michael A. Brennan
Joseph P. Ficalora
Herendino (Dino) Hernandez
Roger L. Hinkley
Daniel M. Kutz
Kristine K. Nissen
Randy Perry
Richard R. Scott
Debby A. Sollenberger
Dr. Ian Wedgwood

Peak Performance Consulting Corporation

Joyce A. Friel

Instantis, Inc.

For more about each contributing author, please see Appendix B, "About the Contributors," in the back of this book.

PART I
PRE-LAUNCH

1 Six Sigma—The Initiative, the Deployment, the Technology

As Larry Bossidy, with whom I worked while at AlliedSignal (now Honeywell), points out in his excellent book, *Confronting Reality* (with Ram Charan), one of the most important competencies an organization can develop is that of driving change. Most organizations find leading change is a very difficult proposition. Larry asserts that, for a company to become good at driving change, it must train itself to drive change.

The change initiative is the building block to driving change. Your company will be reinvented one initiative at a time if each initiative is launched successfully. Larry Bossidy, while CEO of AlliedSignal, preceded the Six Sigma initiative with two smaller initiatives: Total Quality Management (TQM) and Total Quality for Speed (TQS). Using these first two initiatives, he taught the $14 billion company how to launch initiatives. But even Larry would probably admit that the Six Sigma launch was the quickest and deepest of the initiatives launched in AlliedSignal at the time.

The purpose of this chapter is to provide you with an overview of the different dimensions of Six Sigma. I will present the following dimensions:

- Six Sigma—the initiative
- Six Sigma—the alignment
- Six Sigma—the discipline
- Six Sigma—leadership development
- Six Sigma—the methodology
- The Six Sigma playbook

I assume you are interested in deploying Six Sigma for a number of reasons. First, you are interested—perhaps because you are the senior leader—in launching a Six Sigma program to transform your organization. Or, maybe you have recently been chosen to lead this mysterious program for your company, and you want to know more about it. You may want to sell the idea of launching Six Sigma to your senior leadership team and you need more specifics. You've read about Six Sigma and talked to peers in other companies about it. You've got a lot of unanswered questions to be answered before you can feel comfortable. You want to understand the resources required, your time requirements, and the cost of the program. You want to clarify the milestones and actions necessary to produce a model program launch.

I will assume you have a reasonably good strategy. You've surrounded yourself with good people and you have a good instinct about what needs to get done. But you can't quite get your organization to turn the corner. This book is meant for you. Launching a Six Sigma initiative will serve as the impetus your company needs to start the journey from being good to being great. But, even more importantly, following this roadmap to launching Six Sigma in 90 days will ensure a very quick implementation with a 99.99996 percent (i.e., Six Sigma accuracy) chance for success.

And, finally, you want a forecast of the potential measurable impact on your organization's growth and productivity. You also understand that undertaking a major initiative aimed at redefining your organization is a very risky business. You certainly don't want to be known as a leader who has produced yet another program of the month (though every program I've seen has lasted more than a month—shoot, sometimes up to six months).

You are an organizational leader who struggles to move your organization to a new level of performance. Your company's productivity does not represent the typical productivity in your markets and is not where it needs to be. To add to all that, you're not growing fast enough. You are facing challenges ranging from global competition in the corporate arena to greatly reduced government support in the nonprofit arena. Your organization does many things well, but there seems to be something missing on which you can't quite put your finger.

Understanding that you're paid to make money for your company, defeat your competitors, and stimulate them to play in other markets, every good leader dreams of creating a culture that will do just that. You probably did not rise to where you are in your organization by allowing your competitors to hammer you day in and day out. The global rate of economic change is so intense that you feel you must prepare your organization to quickly recognize market changes and react quickly to invent new business models to achieve competitive advantage.

In the heat and pressure of competitive change, you must build an organization that will drive change quickly. You would love to create a new core competency that would allow your company to quickly invent and execute new business models. Because competition is hot and winning is important to the livelihood of all your employees, I have created a playbook with which you can launch a major performance-enhancing initiative—Six Sigma.

But, at the end of the day, your vision embodies the idea of leaving a legacy of an organization where

- Every employee understands the company's business, goals, and vision.
- Every employee knows how he or she contributes to the company.
- Every employee knows how to improve their processes.
- Every employee knows how to solve problems.
- Every function works together seamlessly.

This book is based on over 17 years of direct Six Sigma experience and work with over 45 corporations over the last 8 years. We have experienced the entire continuum ranging from world-class Six Sigma launches (AlliedSignal, 3M, Cummins, and Celanese) to launches bordering on the mediocre.

Many Dimensions of Six Sigma

Six Sigma has many faces. Surprisingly, little of the benefits of Six Sigma have to do with the statistical techniques that are often associated with it. While in its simplest definition—a methodology that focuses on processes to improve growth and productivity—Six Sigma is much richer than a set of advanced statistical tools. Let's look at the value proposition for Six Sigma from your point of view, that of a leader. I will address Six Sigma as a

- Change initiative.
- Method of aligning actions to strategy.
- Driver for operational discipline.
- Leadership development program.
- Methodology.

Six Sigma—The Initiative

Even with the success of the Six Sigma launch, the CEO, Larry Bossidy, kept the pressure on AlliedSignal to change by launching a technology initiative (Technology Excellence), an initiative to connect to our customers (Customer Excellence), and an initiative to reduce paperwork, a Digitation initiative. So, playing on the success of previous initiatives, AlliedSignal developed a core competency to drive change. This book uses that learning cycle as the foundation of the launch methodology.

Because Six Sigma has been so successful in accelerating the performance of so many companies, this book focuses on the process of launching the Six Sigma initiative within a very fast 90 days. The evidence clearly shows that many senior leaders who departed Six Sigma companies and who have moved to other companies have introduced Six Sigma as one of their first change initiatives. Fred Poses (American Standard), Jim McNerney (3M), David Weidman (Celanese), Jim Sierk (Iomega), Paul Norris (WR Grace), Bob Nordelli (Home Depot), Wes Lucas (Sun Chemical), Dan Burhnam (Raytheon), and Ed Breen (Tyco) all left the Six Sigma companies AlliedSignal, GE, and Motorola to drive Six Sigma early in their new assignments—and successfully, I might add.

By learning how to launch the Six Sigma initiative, you will also learn how to launch any initiative quickly and efficiently. You will launch Six Sigma using the same milestones required to launch any initiative. In fact, our launch methodology follows John Kotter's of Harvard University change model, as described in his fine book, *Leading Change*. This model will be explained in further detail in a later chapter.

Therefore, you will have a well-documented theoretical framework within which to launch follow-on initiatives. You will successfully launch Six Sigma, and you will gain fuel-injected performance as a result. As a result, you will achieve the twofold benefit of earning the impressive results of Six Sigma and completing a learning cycle in launching successful initiatives. Success in your Six Sigma launch will ensure that your organization will have the core competency to embark upon future change initiatives with confidence and speed.

Six Sigma—The Alignment

Six Sigma allows an organization to align its processes, people, and strategy with the economic market (voice of the customer). This alignment is not easy to attain, but is imperative for success in today's marketplace. In Stephen Covey's recent book, *The 8th Habit*, he reports the results of a Harris Interactive poll addressing an organization's ability to focus and execute their highest priorities. Some 23,000 employees were polled, with some surprising results:

- Only 37% said they have a clear understanding of what their organizations are trying to achieve and why.
- Only 1 in 5 are enthusiastic about organizational goals.
- Only 1 in 5 said they had a clear line of sight between their tasks and their organizational goals.
- Only 15% felt their organization fully enables them to execute their goals.
- Only 10% felt their organization holds people accountable for results.

My consulting practice has worked with over 50 corporations in launching Six Sigma. I would say the preceding results were typical of most companies before Six Sigma was launched. I will show in Chapter 5, "Strategy: The Alignment of External Realities, Setting Measurable Goals, and Internal Actions," how to use Six Sigma to align your organizational activities to your strategies, people, and processes with your company's external realities and financial targets. A sound Six Sigma launch will be the first step to improving the connection between your people, their actions, and their impact on your company's performance.

Six Sigma will establish a clear link among the Six Sigma projects you identify each year with the external realities of your business, your strategic objectives, and metrics. Your organization's population will view with clarity how their actions impact their organization's performance, from the lowest level to the executive office. Six Sigma will be the tool with which you bring your strategy to reality—one Six Sigma project at a time.

SIX SIGMA—THE DISCIPLINE

Six Sigma provides one very clear benefit within the scope of process improvement—the creation of a disciplined organization. This does not mean you will create a bureaucracy to support Six Sigma. You will, however, create a straightforward, comprehensive, and comprehensible system by which strategic projects are identified, prioritized, resourced, tracked, and completed. While you will create an infrastructure and systems to support your Six Sigma program, this infrastructure will not in any way obstruct your organization from becoming excellent in business execution.

As Jim Collins says in his book, *Good to Great*, "Sustained great results depend upon building a culture full of self-disciplined people who take disciplined action." The goal is to create a culture of discipline with an ethic of entrepreneurship. The infrastructure you create to support Six Sigma will ensure you are getting the return on investment in dramatic business results that you deserve.

By training your people to follow the Six Sigma process improvement methodologies, you will create a disciplined organization around problem solving and process improvement. This strategy makes sense because everything your customers see from you of value is the output of a set of business processes. You will see consistency in the way your organizations select, prioritize, resource, and complete strategic projects. You will see projects that are consistently aligned to your strategic goals. You will create a new core of process improvement experts—Master Black Belts, Black Belts, and Green Belts. And you will see serious accountability for results. With all that, you will see enormous growth in creativity and innovation. You will also allow for each of your businesses to apply Six Sigma to the problems that are unique to them.

While the Six Sigma methodologies provide a standard, disciplined roadmap for attacking problems, the tools within the roadmaps provide the insights that stimulate creative solutions. With the Six Sigma enterprise support systems you implement, you will be able to track the impact of your new Six Sigma capabilities to the bottom line in terms of earning per share, productivity improvements, and organic growth.

SIX SIGMA—LEADERSHIP DEVELOPMENT

Smart leaders are constantly trying to surround themselves with the right people. You will find that Six Sigma is a great developmental process for your future leaders. James Kouzes and Barry Posner in their book, *The Leadership Challenge*, say great leaders

- Challenge the process.
- Inspire a shared vision.
- Enable others to act.
- Model the way.
- Encourage the heart.

Challenge the Process. You will train Master Black Belts, Black Belts, and Green Belts—hereafter referred to as Belts—to challenge (and improve) the process by the very nature of Six Sigma. Any shortfall in performance can always be tracked back to a poorly executed business process. Six Sigma provides a set of roadmaps that specifically delineate the steps with which to challenge any process.

Inspire a Shared Vision. You will also train your Six Sigma Belts to become dynamic team leaders. They will learn the value of inspiring a shared vision with respect to the projects they are assigned. They will enable others to take action by teaching their teams the tools of Six Sigma.

Enable Others to Act. Each person on a Six Sigma team will be able to walk away and solve some of their own problems. Your Six Sigma Belts will model the way by leading their teams from the front to successfully solve problems. They will always be the leader their team looks to for guidance. As they progress through their career ladders, as midlevel and senior-level leaders, they will set clear expectations because it will be obvious they have already done Six Sigma and they know what they can get out of Six Sigma.

For example, while working with Jim McNerney of 3M in launching Six Sigma, Jim drove one of the fastest launches of Six Sigma on record. But it worked because it was obvious when he spoke that he knew what he was talking about. In fact, he taught me a thing or two about Six Sigma. He had launched Six Sigma into two divisions of General Electric successfully and it showed.

Model the Way. Six Sigma provides the perfect leadership training experience. The Six Sigma Belts demonstrate their capability in leading through well-defined, well-resourced projects with quantitative goals for accountability. The good leaders will be easy to identify. Poor leaders will become evident as well. You will rely on a quantitative measure to select your high-potential people during your succession planning—by tracking how much money they brought to the bottom line in their Six Sigma projects.

Encourage the Heart. And, finally, your Six Sigma Belts will know how to encourage the heart. They will show their teams that they can win against the odds. They also will show their teams that they can accomplish amazing things if they keep the goal in sight and follow the discipline of Six Sigma roadmaps.

Every company has a method to identify and groom high-potential employees. The problem with the identification process is the lack of quantifiable qualifications. There is error in the selection process no matter how good it is. Real A-players are missed and lower potentials are sometimes selected. With Six Sigma, it's easy. If a project leader brings in $1 million to the bottom line, chances are good that he or she has a lot of potential. You will find quite a few "sleepers"—high-potential people who have been flying under the radar, but who flourish within the discipline of Six Sigma.

A senior leader in Cummins said it best, "All our future leaders need to be experts in process improvement." Think of it—an outstanding group of young leaders who are not afraid of breaking paradigms, love driving change, and are experts in process improvement.

SIX SIGMA—THE METHODOLOGY

So what's involved with launching a Six Sigma initiative and what are the methodologies? This section presents a very high-level overview of the Six Sigma initiative. The following chapters will overlay the details that are necessary for a successful Six Sigma launch. The high-level steps in the Six Sigma methodology will include: (1) selecting the right projects; (2) roles and responsibilities of the primary players—Belts and Champions; (3) the right roadmaps and tools; and (4) the right results.

SELECTING THE RIGHT PROJECTS

Selecting the right projects sounds easy at first. The objective is to identify the set of projects that—if completed—will yield the most significant impact of growth and productivity. A good business strategy is the prerequisite to the process of selecting projects. Metrics (operational and financial) for measuring strategic progress is also a requirement. It turns out that the greatest challenge in the Six Sigma methodology is selecting and prioritizing the right projects. I will discuss this challenge in detail in Chapter 9, "Committing to Project Selection, Prioritization, and Chartering," but projects should reflect activities directly tied to either long-term business strategy (known as top-down projects) or short-term business results (known as bottom-up projects). Successful Six Sigma companies have created a disciplined process to select, prioritize, and scope projects. These companies create a vibrant dynamic between senior leadership driving for strategically important projects and everyone else driving for projects that will reduce immediate costs while improving quality or capacity.

Manufacturing companies select their Six Sigma projects around the general metrics of cost, quality, and capacity. Companies representing the service industries will likely be focused on speed and accuracy. Understanding how each metric relates to the business drives the prioritization process. If a business is capacity constrained, for example, the business will focus on improving capacity metrics.

ROLES AND RESPONSIBILITIES OF THE PRIMARY PLAYERS—BELTS AND CHAMPIONS

Six Sigma initiatives require the selection of people representing several roles in process improvement. These people lead and support Six Sigma projects and will become the process improvement engine of the organization. The roles—*Champions* (Initiative, Deployment, and Project), *Master Black Belts*, *Black Belts*, and *Green Belts*—have distinct responsibilities in driving Six Sigma.

Two types of champions provide protective support for the Six Sigma program. The *Initiative Champion,* sometimes known as a sponsor, ensures that the program is institutionalized at the organizational level. He or she is accountable for measurable business results to a fairly high level in the organization. The Initiative Champions will reside at the Director or Vice-President level within the organizational structure. Many times, business leaders double as Initiative Champions.

The *Project Champions* work directly with the Belts to ensure that the Belts have the proper resources and organizational support to successfully complete their projects. Project Champions have the most critical role in a Six Sigma launch. They are most accountable for the successful completion of the company's portfolio of Six Sigma projects. An executive at Cummins said, "There is no such thing as an unsuccessful Black Belt, just unsuccessful Champions."

Companies internalize (i.e., become consultant-free) their Six Sigma programs by developing *Master Black Belts* (MBBs). The MBBs are expert in the process improvement roadmaps and tools. They will also be proficient in driving deployment strategies. They will ultimately replace external consultants and will provide mentoring and training required for the long-term success of the Six Sigma program. MBBs create value through the mentoring process (i.e., ensuring that Black Belts and Green Belts are successful) and secondarily through training new Belts in process-improvement skills.

For example, one of AlliedSignal's first Master Black Belts, Bill Hill, mentored 12 Black Belts who achieved some $30 million total for their first 12 projects. Clearly, Bill was a success in ensuring that each project was completed as quickly as possible. MBBs also work closely with the corporate leadership in project selection and Six Sigma deployment. The MBB is the Champion of the technical side of Six Sigma. They are experts in all the tools included in the number of Six Sigma roadmaps that are available.

Black Belts (BBs) drive strategic (i.e., big and important) projects; companies consider Black Belts as strategic resources. They receive extensive training in the roadmap and tools discussed in later chapters and demonstrate clear expertise in solving complex chronic problems. Our successful clients allow their Black Belts to work full-time on their projects. The standard procedure within a Six Sigma deployment expects Black Belts, after they learn to use one of the several process improvement roadmaps, to drive two or three project teams simultaneously.

Green Belts (GBs) drive shorter-term tactical projects. Green Belts are widespread throughout the company, and use one of the process improvement roadmaps to solve problems within their area of the company. Green Belts are generally part-time on projects and carry out their usual responsibilities. The training of Green Belts is 50% or less of the training of a Black Belt, and they can be trained by Black Belts or Master Black Belts. Therefore, the Green Belt training program can be widespread and delivered locally and cost effectively. Green Belts always attend training with an important project.

THE RIGHT ROADMAP AND TOOLS

Most companies do a good job of identifying important problems, identifying a team leader, and pulling the right team together. The teams, however, usually hit a wall when actually trying to solve a problem. Companies generally do not have a recognized, standardized, and systematic way of solving chronic problems. Companies now move away from relying on the top 10% (high-potential) employees to solve big problems and turn to the top 40% to solve problems. Six Sigma provides this inherent power.

Having common roadmaps and tool sets provides the problem-solving backbone for the company that yields a common language and set of expectations. For large, complex corporations such as Motorola, Honeywell, and GE, this common language ties the corporation together and also provides a way to easily assimilate new acquisitions.

THE RIGHT RESULTS

The final evaluation of the success of any Six Sigma program is the verification of quantifiable business results, usually measured in pretax income per annum. My favorite metric is dollars per trained Belt. Larger companies train hundreds of Black Belts and Green Belts, who are completing projects all around the company. In the process, companies invest significant resources and training to support their Six Sigma efforts. Companies successfully implementing Six Sigma expect a return-on-investment (ROI) of 30x to 100x over a two-year period.

Organizations will normally implement a centralized project tracking system to determine the effectiveness of the Six Sigma program. By systematically tracking project results, the program is constantly validated and improved. If done correctly, the financial analysts of the organization will be able to convert project results to earnings per share.

THE PLAYBOOK

As a leader, you face the ultimate challenge of driving dynamic change across your organization, resulting in the organization's long-term success. Virtually every organization—profit versus nonprofit; manufacturing versus services—meets significantly new challenges every day. These challenges attack you from every direction, both domestically and globally. As you make your way through the "fog of war," you try to determine the actions that, if executed properly, will transform your company into a high-performing engine. In short, you have a vision of what needs to get done, but you may not know exactly how to do it.

This is the playbook for launching a Six Sigma initiative in your organization. I will present a play-by-play plan to launch Six Sigma in 90 days. I define the first 90 days as the time interval starting with the first executive training session and ending with the first day of the first training wave of Six Sigma Black Belts (Black Belts will be your future process improvement leaders). This playbook is organized into three parts: (1) Pre-launch preparation; (2) Launch; and (3) Post-launch actions.

PART I: PRE-LAUNCH

The pre-launch preparation chapters provide you with the specific actions and milestones necessary to get to the starting line.

Chapter 2, "The True Nature of Six Sigma: The Business Model," provides you with insights into the process of focusing on money (e.g., results), finding the money, and delivering the money. I also address the business model to ensure alignment between your Six Sigma activities and your balance sheet.

Chapter 3, "Six Sigma Launch Philosophy," gives you an executive overview summarizing the nature of a typical Six Sigma deployment. My team and I have sifted through some 50 Six Sigma launches to find the best practices.

Chapter 4, "Getting Early Support: Selecting a Six Sigma Provider," supplies you with the rationale for using an external support group and a list of requisites to consider when evaluating providers. I also provide a quantitative decision matrix and a draft of a request for proposal.

Chapter 5, "Strategy: The Alignment of External Realities, Setting Measurable Goals, and Internal Actions," delineates the process by which projects are linked to the organizational strategy. This chapter describes the process by which projects are selected and prioritized to obtain a direct line of site to the strategy.

Chapter 6, "Defining the Six Sigma Program Expectations and Metrics," discusses how to measure the success of your Six Sigma launch with metrics that are directly tied to your strategic and annual operating plans. This chapter instructs you in how to develop your strategic metrics and, more importantly, how to develop aggressive but achievable goals.

PART II: THE FIRST 90 DAYS

This part covers the time from the initial executive training workshop to the launch of the first wave (class) of Belt training. This part provides specific actions and milestones that lead to a successful Six Sigma launch. Each chapter provides a step-by-step roadmap to accomplish a specific set of actions and provides recommended timing. Launching a

Six Sigma program in 90 days is realistic. Very large companies such as Honeywell and 3M have done so, along with several smaller companies. In fact, 3M beat the 90-day timeframe by about 60 days.

Chapter 7, "Defining the Six Sigma Project Scope," helps you decide what your Six Sigma deployment will look like. The chapter recommends how you tie Six Sigma to earlier initiatives, and compares the advantages and disadvantages of pilot projects versus full-scale organization-wide deployments. I discuss approaches that consider division to division and geographic deployments. In addition, the chapter reviews the multitude of Six Sigma and Lean programs that are available to you to deploy in your organization. Depending on your strategic requirements, this chapter advises you on which programs to launch and in what order.

Chapter 8, "Defining the Six Sigma Infrastructure," outlines the roles and responsibilities of the key players in a Six Sigma launch. It will offer selection criteria and timing. Data infrastructures are also addressed.

Chapter 9, "Committing to Project Selection, Prioritization, and Chartering," is one of the most important chapters in the book. This chapter provides a step-by-step methodology by which breakthrough projects are identified, prioritized, and chartered. By following the suggested methodology, you can learn the valuable core competency of strategic project selection.

Chapter 10, "Creating Six Sigma Executive and Leadership Workshops," recommends how these very critical workshops should be developed and formatted. This chapter presents several different approaches to each workshop, along with sample agendas. I discuss best practices as well. These workshops set the stage for the next step, which is to develop an overall, long-term deployment plan.

Chapter 11, "Selecting and Training the Right People," reviews the roles of the significant Six Sigma players and details the training required to fulfill these roles. The chapter also discusses the pros and cons of hiring personnel externally and developing personnel internally. And, finally, the chapter provides insights into the process of internalizing the Six Sigma training.

Chapter 12, "Communicating the Six Sigma Program Expectations and Metrics," provides ideas for developing a comprehensive communications plan. Emphasizing the who, what, when, where, why, how, and how much, this chapter discusses the coordination of various communication channels. This chapter provides you with the essence of marketing Six Sigma throughout your organization.

PART III: POST-LAUNCH

This part discusses the means and methods for institutionalizing your Six Sigma program for the long term. We provide a leadership roadmap, along with specific actions that must occur every year for the program to last. Milestones like compensation plans, career ladders, and software support are detailed.

Chapter 13, "Creating the Human Resources Alignment," provides guidance in aligning Human Resource systems to the Six Sigma program. This chapter discusses organization structure, succession planning, career planning, measuring Belt performance, and reward and recognition.

Chapter 14, "Defining the Software Infrastructure: Tracking the Program and Projects," provides an overview of the requirements of a project-tracking software system. The chapter also provides some examples of systems I have encountered over the years.

Chapter 15, "Leading Six Sigma for the Long Term," provides a detailed leadership roadmap that, if followed, ensures the long-term success of your Six Sigma program. This step-by-step approach is based on our work with over 40 companies over the last 15 years.

Chapter 16, "Reinvigorating Your Six Sigma Program," provides methods to assess your current deployment and reinvigorating it based on identified gaps in the deployment. Success in deploying Six Sigma provides the opportunity to launch successive change initiatives.

This playbook is designed to provide you with a turnkey method to launch a Six Sigma initiative in 90 days. By using this fast approach, you can quickly and effectively align your people to your strategy, processes, and customers. By doing it within 90 days, you ensure that you will have a different company within 12 months. These fast launches also demonstrate (require) the full commitment of your leadership team. These 90 days will be the most memorable of your career.

2 The True Nature of Six Sigma: The Business Model

With Randy Perry

This chapter will cover the inherent nature of Six Sigma and most of the necessary elements of an effective Six Sigma launch at a business level. Although Six Sigma has a reputation of being a manufacturing program, it is much more complex. Essentially, Six Sigma is about streamlining and money. This chapter is aimed directly at you, a leader who wants to dramatically impact the performance of his or her organization very quickly. This chapter will help you clarify the potential business impact of Six Sigma and will give you a general sense of what needs to be done to launch Six Sigma quickly and effectively. The elements and methodologies presented in the chapter will be covered in more detail in later chapters.

While focusing on a fast Six Sigma launch, however, we must never lose sight that the real purpose of your organization is to make money.

The Business Model

How can you make a lot of money consistently over the long term? Now that is the question! The answer may lie with one of the most effective CEOs of our day, Larry Bossidy. Larry Bossidy (with Ram Charan) in his book, *Confronting Reality*, presents a unique business model that links four components:

- External realities
- Financial targets
- Internal activities

These three components are linked into a dynamic system that produces new business models with which to move your business smoothly into new markets and new products. Bossidy and Charan state, "Linking and iterating the financial targets, external realities, and internal activities, and searching for the right mix in each of the three components of the business model determines the accuracy of the final product." You will make money by effectively linking your internal activities to the external realities and setting appropriate financial targets.

This fresh approach to facing the realities of today's competitive marketplace happens to represent our approach to Six Sigma. It's no coincidence that, in 1994, Larry led the best launch of a Six Sigma initiative since Motorola's launch in 1987 and that I base this book largely on that deployment. This chapter presents the business case for Six Sigma, and we'll use Larry's model as a basis. You will see how Six Sigma allows you to effectively link the three components of the Bossidy model to ensure that you hit your financial targets year after year.

I met with a new president of a chemical company to talk with his leadership team about the benefits of Six Sigma. He had recently arrived from AlliedSignal, and he explained to his leadership team, "Before Six Sigma, we developed our annual operating plan and then executed the best we could, but hoped at the end of the year we would have met our goals. We deployed Six Sigma, and I not only didn't have to hope to meet our goals, but I sometimes got up to 2x my usual end-of-the-year bonus." In fact, in a classic case study, 3M's CEO, Jim McNerney, used Six Sigma as one of his first steps to linking the three components soon after he arrived at 3M.

Assessing External Realities

First, let's look a little closer at Larry's model. The first component, assessing the external realities, looks at the business and economic environment in which you currently compete. You will assess four elements, as follows:

- The broad business environment
- The financial history of your industry
- Your customer base
- Root-cause analysis

In reality, these four assessment elements provide the inputs of systematically collected market, industry, and customer data (with the attractive option of using Six Sigma–based marketing tools and roadmaps) and the output of a new business model based on intuition and judgment. Because you are deep into the complexities of the marketplace, there is no guarantee that the models you develop are the right ones with which to move forward. But by assessing the external realities systematically and iteratively, you greatly improve the probabilities that your leadership will understand what your company needs to do to be competitive.

SETTING FINANCIAL TARGETS

Once you determine what needs to get done strategically, you then set financial targets linked to the external realities. These targets include a subset of the usual suspects in the financial world:

- Operating margins
- Cash flow
- Return on capital
- Revenue
- Return on investment

The financial targets you set will depend upon your external realities. For example, a company might demonstrate mediocre operating capability, so financial metrics addressing productivity may have more aggressive goals than those addressing growth. You might set 20% improvement targets on operating margins and cash flow and 10% improvement targets on revenue.

Once financial targets are identified and goals set, Six Sigma really starts. Clear measurable metrics and targets represent the foundation of Six Sigma. The old adage, "what gets measured gets done," applies here. With financial metrics driving your Six Sigma program, all activities, if prioritized properly, will directly affect your company's performance.

THE NATURE OF SIX SIGMA

Six Sigma is focused on process improvement. Anything your customers see from your company are the outputs of a set of business processes. In fact, you could say that a Six Sigma company views poor results as symptoms of poorly designed processes that produce process errors. Six Sigma methodologies provide you the quantitative understanding of the relationship between the process outputs and the process inputs. The basic formula is simple: The output of a process is a function of a set of the inputs of a process (Y = f(x's)). But of course, this is Six Sigma, so I have to produce some kind of formula. So we'll use:

$$Y = f(x_1, x_2, \ldots, x_k)$$

Where Y is the output and the Xs are the inputs. We would say, "Y is a function of the Xs."

We would call our selected strategic financial metrics (e.g., operating margins, cash flow, revenue) our "Critical Ys." Examples of some of the output/input functions would be as follows:

- $Y_{(\text{operating margins})} = f(\text{revenue, product costs, business costs})$
- $Y_{(\text{cash flow})} = f(\text{profits, working capital})$
- $Y_{(\text{return on capital})} = f(\text{volume, average selling price, discounts})$
- $Y_{(\text{revenue})} = f(\text{volume, average selling price, discounts})$
- $Y_{(\text{return on investment})} = f(\text{profit, investments, net assets, asset turnover})$

You may argue with the inputs, but you get the point. Every financial metric is driven by a set of inputs, and these inputs can usually be tracked back to operational processes. We'll refer to each of the preceding Ys as Critical Ys because they are financial metrics that drive the success of the company. If you understand and control the variation of the inputs, you control the variation of your Critical Ys (outputs).

Let's take one of the preceding financial metrics and move a little farther down the Six Sigma path. Let's pick the one that is near and dear to our hearts, cash flow. For the process output, or our Critical Y—cash flow—we might have an equation like $Y_{(\text{cash flow})} = f\,(\text{profits, working capital})$. The output, cash flow, the Critical Y, is a function of the inputs, profits, and working capital.

The next step is to consider the inputs of profits and working capital as second-level outputs, or Little Ys. We then identify and prioritize which operational inputs have the most influence on each of profits and working capital. Some example inputs for the $Y_{(\text{profits})}$ would be the following:

- Manufacturing yields
- Cost of poor quality

- Cost of goods sold
- Invoice to cash
- Supplier cost variances

Some example inputs for the $Y_{(working\ capital)}$ might be the following:

- Manufacturing yields
- Product cycle time
- Work in progress levels
- Finished goods inventory
- Raw materials inventory

In the Six Sigma world, we would then analyze the inputs to the process outputs, $Y_{(profits)}$ and $Y_{(working\ capital)}$, and identify which process inputs drive the variation in these two process outputs and optimize them. By understanding and optimizing the inputs, we will then control the variation and optimize our outputs. With the right work in the right places, we will make huge impacts on profits and working capital, which will in turn have a significant and positive impact on cash flow (our Critical Y). Figure 2.1 shows the relationships between the Critical $Y_{(cash\ flow)}$ and the Little Ys (profit and working capital). And finally, I've mapped the Little Xs into the Little Ys. The point of all this? Well, we see several operational inputs we might address to improve cash flow. We notice that if we improve manufacturing yield, we would improve both Little Ys of profit and working capital. If we do some baseline measurements, and we find we have a lot of room to improve yields, then a set of Six Sigma projects addressing yield would be valuable to the company. As we improve yield, we improve both profit and working capital and, in turn, we improve cash flow (our strategic financial target).

Likewise, if we find we are inefficient in the way we manage our work in progress and we have a lot of potential to improve there, we would launch a set of projects around improving work in progress depending on Lean manufacturing methods. These projects will be a combination of Lean and Six Sigma techniques. By improving our work in progress levels, we directly impact working capital (our Little Y), which in turn impacts cash flow (our Critical Y).

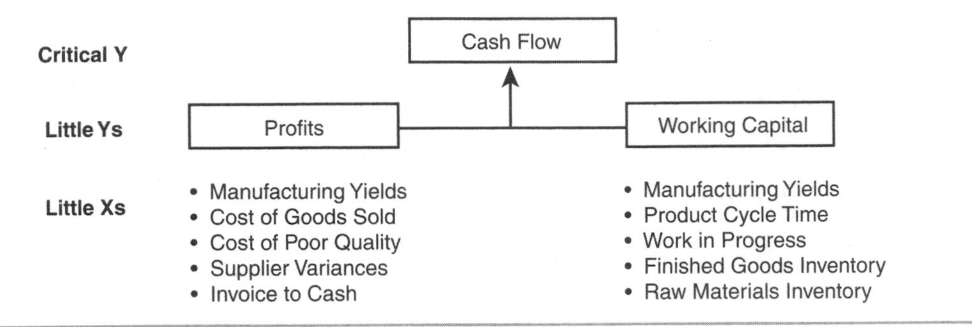

Figure 2.1 The Six Sigma input/output model depicting the Critical Y (strategic financial target), Little Ys (inputs to the Critical Y), and Little Xs (operational targets).

LINKING INTERNAL ACTIVITIES

Bossidy's model includes the third component: internal activities—the scan of the environment. Now that you have assessed the external realities and your gaps in the market and have created a linked set of financial targets, you are ready to execute the new business model by linking your internal activities. The elements of this component encompass the following:

- Strategies
- Operating activities
- People selection and development
- Organization

These internal elements must be clearly linked to the external realities. The strategy is the document that shows the alignment between the external realities, the financial targets, and the set of elements including operating activities, people, and organization.

Given a solid strategy (the prerequisite to launching a successful Six Sigma program), Six Sigma directly impacts operating activities and people selection and development. The companies that have been successful in deploying Six Sigma have been given options to defining new strategies. Operating activities as defined by Larry Bossidy include the initiatives and processes "that enable your business to reach the desired financial targets and execute strategies, such as product launches, sales plans, and measures to improve productivity."

Because Six Sigma was born out of manufacturing in Motorola in 1987, Six Sigma is notorious for streamlining and optimizing manufacturing processes. But Six Sigma also has a fine history in impacting business processes and new product and services development, as well as sales and marketing. In fact, every operating area in your company will have a special Six Sigma roadmap to use to make their part of the company great.

Bossidy suggests, when evaluating your strategy, you answer the questions: "Do you have the right people to pull off the introduction of new initiatives? Are they properly deployed—are the right people in the job?" Six Sigma has long demonstrated its capability to develop the future leadership of companies that use it. In fact, Six Sigma launches in AlliedSignal and General Electric spurred the creation of new corporate positions throughout the corporate world today. There are entire headhunting companies and web sites dedicated to finding Six Sigma Master Black Belts, Black Belts, and Green Belts. These Belts have learned the art and science of process improvement through completing important projects successfully.

Again, Six Sigma is process focused. Your company is defined by hundreds of processes—a legal department alone in a very large company identified over 40 processes they manage. Your customers see only the results of how efficient and effective these processes are. If you are not focused on systematic process improvement, you will ultimately lose in the marketplace to competitors who do. So, to rephrase Larry's business model in terms that are a bit more Six Sigma friendly, we can say that the growth of your company depends on how effectively you link the following:

- Dynamics of your industry
- Corporate strategy and organization
- Policies and procedures
- Product or service technologies
- Operational processes
- Business processes

Six Sigma directly impacts the last three: technologies, operational processes, and business processes. Understanding the dynamics of your industry and developing a corporate strategy and organization to address those dynamics tells your organization what to do. Six Sigma shows your organization how to do it.

So, Six Sigma is like having a corporate personal trainer. Six Sigma will help quantitatively define business goals and achieve them, will bring the necessary discipline to process problems, and will align your company's resources to achieve specific goals and—most importantly—convince people these goals can be achieved.

FOCUSING ON, FINDING, AND DELIVERING THE MONEY

Now that we've defined the business model within which we will work, let's now focus on the real important issue of making money. We will focus on a three-step process:

1. Focusing on the money.
2. Finding the money.
3. Delivering the money.

FOCUSING ON THE MONEY

A Six Sigma program, launched correctly, will provide a discipline throughout the organization around the powerful concept—making money is a key objective. By creating this focus on money, you will see behaviors changing as people consider their actions in relation to what goes to the bottom line of the balance sheet.

FINDING THE MONEY

Six Sigma provides a methodology and toolkit with which to find opportunities for financial improvement. Starting with your financial targets, your organization will identify and prioritize projects that are linked to these targets. The result of this process produces a set of high-impact projects that are clearly linked with your financial targets and ultimately with the external realities.

DELIVERING THE MONEY

Now this is where Six Sigma really shines. Once the set of high-impact projects are identified, they will be chartered and your leadership will assign Black Belts or Green Belts (Belts) to drive these projects to completion. Six Sigma provides these Belts with structured methodologies with which to execute these projects and deliver the financial value to the company.

Your financial people will be heavily involved in validating the financial impact of each completed project, and you will have developed an infrastructure of Champions and Master Black Belts to ensure a high project-completion yield. Statistically valid data will drive the decisions about financial impact to be data based.

To give you an example of the financial impact of a large set of projects, AlliedSignal delivered over $880 million in pretax income within the first two years of the program.

This was based on the work of 1,200 Black Belts and 7,000 Green Belts. That was almost equivalent to adding a $4.5 billion-per-year business to AlliedSignal's portfolio without any of the associated capital.

The engineered materials sector of AlliedSignal—a $4 billion-per-year business—delivered over $200 million in pretax income with a cadre of about 300 Black Belts and 800 Green Belts. The Black Belts were averaging about $350K per project and the Green Belts were averaging about $75K per project.

In the process, AlliedSignal created some 8,000 stars—people who were given great projects with the resources and commitment—who went out and made a huge impact on the company. AlliedSignal gave motivated people the opportunity to release their creative problem-solving potential and allowed them to see the results of their efforts hit directly on the bottom line. One 23-year-old Black Belt delivered 3 cents per share with one project! With these three process steps—focusing on, finding, and delivering the money—let's take a look at each step in more detail.

FOCUSING ON THE MONEY

In recent years, Six Sigma has been associated with quantitative improvements in growth and productivity. But, Six Sigma hasn't always had the reputation. In its early years with Motorola, Six Sigma was primarily associated with systematic improvements in product quality. At this point in our discussion of Six Sigma and the focus on money, a quick review of the evolution of Six Sigma from 1987 to the present would provide some context to the following sections.

SIX SIGMA EVOLUTION

Motorola. Historically, Motorola developed Six Sigma in 1987 to eliminate process defects. As a manufacturing manager with Motorola in those days, I can say we were evangelistic about preventing defects. We operated with the goal of reducing our defects by 68 percent year over year (equating to a 10x improvement every two years). In fact, my defects-per-unit (dpu) trend charts where shown monthly to Motorola corporate operations reviews. The metric and visibility were perfectly clear.

If you were not showing improvement in your dpu trend charts, you had very interesting meetings with the plant manager. However, we rarely evaluated the direct business impact of these defect elimination projects. We just believed in our hearts that removing defects allowed us to produce products better, faster, and cheaper.

That was generally the case because we saw countless dramatic case studies addressing business improvement. But there were instances when we would work on a project that might eliminate defects within a certain process, but these activities would sometimes have little or no effect on overall business.

AlliedSignal and Beyond. Six Sigma changed from the focus on reducing defects to focusing on the money that resulted from the impact of two men, Richard Schroeder and Larry Bossidy. Rich launched Six Sigma for Asea, Brown, Baveri (ABB) in the U.S. in 1992. ABB is a large global company with a strict bottom-line focus. Rich had been a driver of Six Sigma during the Motorola heydays, but learned quickly the usefulness of focusing Six Sigma on bottom-line results.

Rich then moved to AlliedSignal in 1994 to launch Six Sigma for Larry Bossidy. Larry, coming from the numbers-oriented GE, appreciated the focus on the money and drove Six Sigma, with the help of Rich, to ensure a significant impact on earnings per share. Stock prices went from the mid-$30s in 1995 to the mid-$70s in 1997, and AlliedSignal delivered almost $1 billion of pretax income in the first two years. The next thing you know, Larry Bossidy tells his friend Jack Welch, and GE launches a huge initiative. Six Sigma has never been the same.

Six Sigma and TQM. A question about Six Sigma—How is it different from Total Quality Management (TQM)?—pops up frequently. TQM, developed in the 1980s to address quality, led a very large number of companies that were dissatisfied with the results. In fact, surveys done in the late 1980s and early 1990s by Arthur D. Little, McKinsey & Co., Ernst and Young, A.T. Kearney, and Rath and Strong indicated that for every company pleased with its TQM results, two other companies were not. The TQM initiative usually took about three years to launch and still didn't make much difference in organizational performance.

Unfortunately, many companies sponsoring TQM fell into the trap of confusing activity for results. Projects were often selected to drive "Quality for quality's sake." Thousands of people were trained in very basic quality tools and thousands of projects were completed. Award ceremonies were ubiquitous. Larry Bossidy referred to TQM as "mugs and hugs," and didn't want to have anything to do with it.

Although Six Sigma uses many tools common to TQM, it is quite different. Some of the ways in which Six Sigma differs from TQM include the following:

- Linking improvement activities to the external realities, financial targets, and strategy
- Focusing externally on the customer and markets
- Focusing on and delivering money and financial results
- Selecting projects based on delivering financial benefits
- Creating an organization that will solve complex process problems with sophisticated tools

- Providing specialized improvement roadmaps for specialized problems
- Delivering of quick results

One senior leader put it best: "Before TQM, we made poor-quality products that didn't sell. After TQM, we made high-quality products that didn't sell." Just remember Motorola's decision to stay with their high-quality analog cell phones and then watched as the digital market went right by them. Because TQM didn't link activities to the strategy and the critical business processes, it never quite hit the mark.

SIX SIGMA: FOCUS ON MONEY

Six Sigma has money as a primary focus. But consider what that means. Your company can create money in three ways—bottom-line growth (productivity), top-line growth (growth), and freeing up cash. Therefore, all Six Sigma projects should be linked to the organizational strategy and be directed at hitting growth targets, cash targets, and productivity targets.

A comprehensive Six Sigma program includes specialized process improvement roadmaps to be applied to different classes of process problems. Designing a new product or service, for example, requires a radically different approach when compared to optimizing a manufacturing process. Solving problems in transactional processes (finance, human resources, sales) requires a different tool set than solving manufacturing process problems.

For projects directed toward improving productivity, the operations-oriented DMAIC (Define, Measure, Analyze, Improve, and Control) roadmap is used. For projects directed at growth, the roadmaps associated with DFSS (Design for Six Sigma) and the MSS (Marketing for Six Sigma) are used. Figure 2.2 depicts the linkage from the strategic plan to the cluster of projects in growth and productivity, and then to the actual Six Sigma methodologies to apply to the projects.

Six Sigma projects, once identified and prioritized, will be chartered. The charter is a contract between the project manager (Black Belt or Green Belt) and his or her organization. The charter defines time lines, expectations, metrics, and resources, and should have the blessing of a financial analyst. The organizational leadership will evaluate the need for Six Sigma projects to support the strategic plan in the short, medium, and long term. As the business drives for long-term growth, DFSS and MSS become critical methodologies for delivering value and hitting financial growth targets.

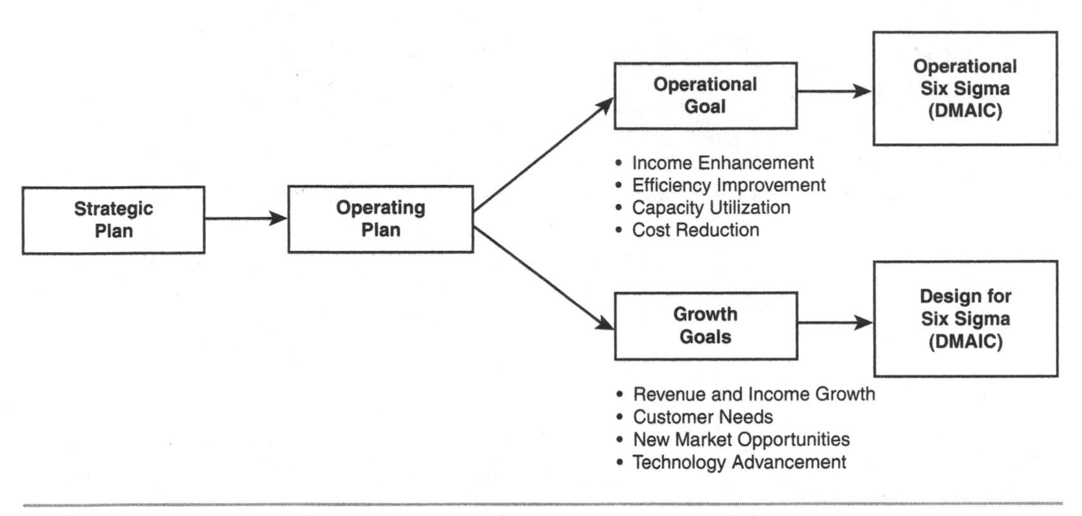

Figure 2.2 How Six Sigma links strategy to methodology through financial targets.

Creating a strong focus on money is the first prelaunch set of activities of a successful Six Sigma launch. Understanding, based on your external realities, where you need to deliver value will drive financial targets and business areas within which you will focus your Six Sigma efforts. The next step is the most difficult step in launching Six Sigma—finding the money.

FINDING THE MONEY

There is money to be made in every corner of your organization. Every function will contribute to productivity, and many functions are involved in driving growth. Your company is comprised of hundreds of processes, each of which creates defects, and adds costs and time due to non-value-added process steps. However, trying to solve all your problems simultaneously will only overload your organization, and you'll end up like the old TQM companies—generating a lot of activity with no results to show for it. The most difficult thing to do in any business is to find the relatively few projects that will generate the biggest results strategically.

Value Mapping. Selecting critical processes, mapping them, and then setting the baseline performance for quality, cost, and cycle time, you will see clearly where you need to work. We'll call this procedure *value mapping*. We saw in a previous section how we move from your strategy to a set of financial targets to potential areas on which to focus. Looking at Figure 2.1, we dissected the financial target of cash flow into several components:

- Manufacturing Yields
- Cost of Goods Sold
- Cost of Poor Quality
- Supplier Variances
- Order to Cash

For this example, we'll pick a process that is linked to Order to Cash—Order to Invoice. Let's look at the value map in Figure 2.3. This might be a typical manufacturing process or a transactional process. I've used the Order to Invoice process as a high-level process for this example. We'll keep things simple and say this process only has three steps. We would send out a team to capture basic performance data (quality, cost, and cycle time).

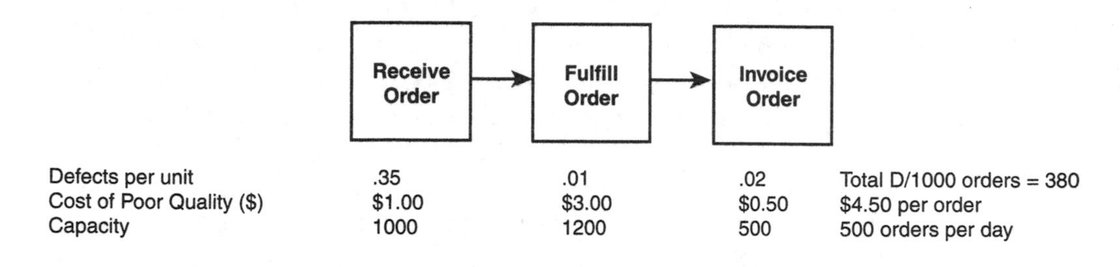

Figure 2.3 Order fulfillment process with three steps showing basic quality, cost, and cycle time measurements.

Now, this is exactly the methodology used with the AlliedSignal engineered materials sector in 1994. Plant managers took two to three weeks to produce the baseline data they needed to complete their value maps. On inspecting the value map in Figure 2.4, we see that the first step, receiving the order, has very poor quality. If this company completed a successful Six Sigma project, it stands to save $150,000 per year.

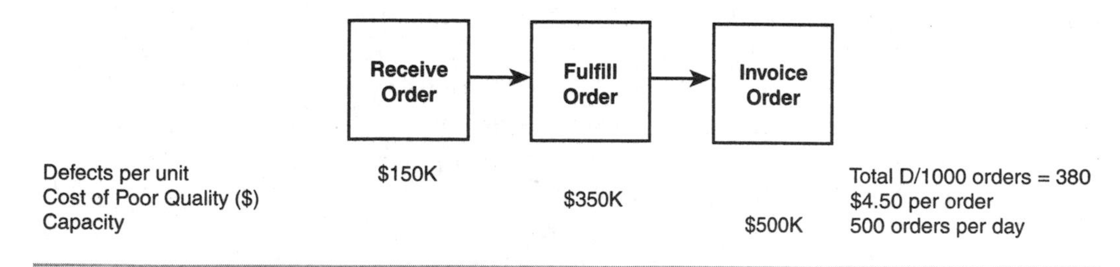

Figure 2.4 Order fulfillment process with three steps showing the estimated financial impact of three projects.

We also see that the second step, fulfilling the order, has the highest cost of poor quality. For every 100 orders, this organization loses $450.00. And, finally, we see that the third step, invoicing the order, has the lowest capacity, with only 500 orders per day. Any improvement in capacity at this step will lead to an overall capacity improvement for the entire process. So, by committing to and completing projects to improve quality in Step 1, cost of poor quality in Step 2, and capacity in Step 3, the company stands to bring some $1,000,000 to the bottom line. These problems have probably existed for years, but they just didn't make the leadership radar screen until they put some performance numbers and financial numbers to them.

So, it would appear that we have three good projects on our hands. However, there is one more step here. The organization will now work closely with their financial analysts to investigate the potential business impact of each project. Figure 2.4 shows the results. The projects attacking cost of poor quality and capacity are estimated to bring $350K and $500K to the bottom line, respectively. These would most surely end up on a Six Sigma project list. The project attacking quality is estimated to bring in only $150K, which is marginal when compared to the other two projects. Depending on resources and other projects identified, the quality project may or may not show up on the final list.

Process Entitlement and Goals. A key concept in driving for aggressive goals turns out to be a concept called *entitlement*. Entitlement represents the best theoretical outcome for any process metric. So, in our preceding example, the process entitlement for defects-per-unit would be theoretically zero. The entitlement for cost of poor quality would be very close to zero—there may be some cost built in for destructive testing and such. The entitlement for capacity would depend on assessing the total value-added time versus the non-value-added time, but would probably be much greater than 1,200 orders per day—let's say for the sake of argument, the entitlement for capacity is 2,000 orders per day.

To set Six Sigma goals, an aggressive approach is to move the process from baseline performance to 50% closer to the entitlement. For example, Table 2.1 shows the baselines and entitlements for the metrics in our invoicing process. The defects-per-unit goal would be halfway between the baseline, 38/100, and entitlement, 0/100. The goal calculates to be 19/100. You know the project addressing defects-per-order is done when 19/100 defects per order is accomplished. Figure 2.5 shows the concepts of baseline, entitlement, and goal setting. This figure shows the difference between a Six Sigma goal—moving quickly toward entitlement—and the usual corporate mode of goal setting—looking at a 10%–20% improvement.

Table 2.1 Chart Showing Process Baselines, Entitlements, and Goals for the Invoicing Process

Metric	Baseline	Entitlement	Goal	Value
Defects per unit	38%	0%	19%	$750K
Cost of poor quality	$4.50	$0.50	$2.50	$540K
Capacity	500	2,000	1,250	$1.6M

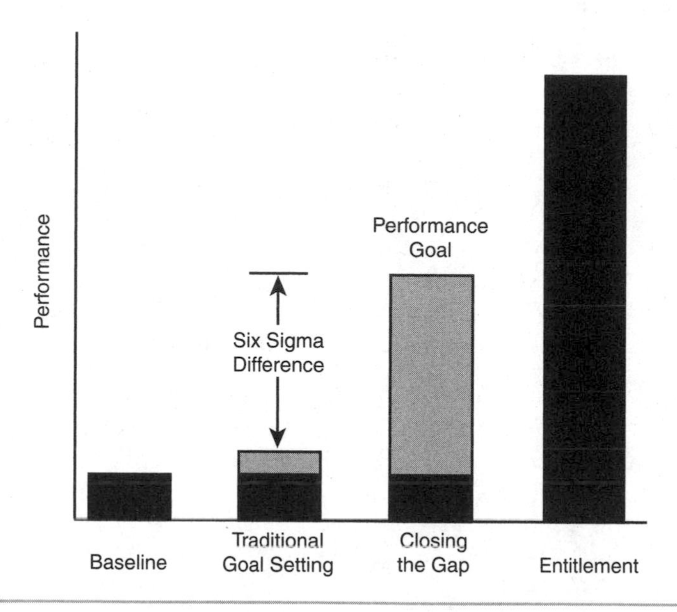

Figure 2.5 Process goal setting using baseline and entitlement. The goal closes the gap between baseline and entitlement by 50%.

The concept of entitlement is valuable psychologically as well as for goal setting. Entitlement allows your leadership team and you to see clearly what you should be able to accomplish as you move toward perfection. Fred Poses, CEO of American Standard, was instrumental in launching a great Six Sigma program into AlliedSignal's engineered materials sector. He told his leadership team at his new company, American Standard, that the most important concept in Six Sigma to him was entitlement.

Measuring the gap between current performance and entitlement allows your company to realize the possibilities for process improvement. It has been said that, "Knowledge of the possibilities is the first step to passion." Understanding clearly how good your company could be is the spark with which you will ignite a terrific launch of Six Sigma.

This concept actually applies to golf. How can a sport that's so frustrating be so popular around the world? I'm convinced that golf's popularity is related to the idea that each golfer catches a periodic glimpse of his or her entitlement in playing golf. That one excellent drive, the miraculous chip shot, and the long putt all provide reinforcement to the golfer on just how good he or she could be. He or she becomes a passionate golfer even though the actual performance isn't close to the entitlement.

ANALYZING PROFITS

Everyone in any business finds profit margins very interesting. Dissecting profit margins analytically produces some great places to find money. We can look at profit quantitatively with this simple formula:

$$\text{Profit} = \text{Revenue} - \text{Cost}$$

That's simple enough. But we can take this equation one step further to gain additional insights into where to look for money:

$$\text{Profit} = [\text{Volume} \times (\text{Price} - \text{Variable Cost/Yield}) - \text{Fixed Cost}] \times (1 - \text{Tax Rate})$$

This formula is more complex, but captures the variables that are heavily influenced by variability—variable cost, yield, and fixed cost. As variability is removed from business processes, those three variables improve and directly affect profit. A more effective system will positively improve price and volume because of improvements in quality and functionality of products or services.

So, projects can come from some of the following areas:

- Increasing sales volume
- Increasing sales price
- Reducing variable cost of production or operations
- Improving process yield
- Reducing fixed cost
- Reducing tax rate

Looking at the preceding, Six Sigma projects can therefore be found in manufacturing, technology, marketing, sales, administration, support, and all other areas of the business. Some great examples of Six Sigma project clusters include the following:

- **Gross sales**
  - Reduce sales price errors
  - Improve capacity productivity
  - Optimize mix of products sold
  - Increase price through good product design
- **Discounts**
  - Reduce special sales allowances
  - Reduce contract discounts
- **Cost of sales (material, labor, overhead)**
  - Improve product flow
  - Lower supplier costs
  - Improve quality in production
  - Improve quality in suppliers
  - Improve payables terms
  - Reduce inventory owned
  - Redeploy assets
  - Improve equipment uptime
- **Warranty**
  - Improve manufacturing processes
  - Improve source of components
  - Reduce testing
  - Streamline warranty process
  - Improve warranty forecasting
  - Eliminate abuses of warranty system
  - Improve design process to improve process and product quality
  - Fix issues quickly to satisfy customers
- **Engineering expenses**
  - Reduce time to develop a product or service
  - Reduce required product testing
  - Improve cycle time of engineering change notices without impacting quality
  - Better utilize partners and suppliers for better designs
  - Improve technology development

- **Marketing expenses**
  - Reduce discounts
  - Eliminate low or no margin businesses
  - Improve capacity productivity
  - Improve processes for communicating product or service volumes
  - Improve forecasting process
  - Improve voice of the customer data collection
- **Administrative expenses**
  - Improve administrative flows
  - Reduce defects in administrative processes
  - Reduce depreciation, taxes, and insurance
  - Eliminate central inventory holdings
  - Reduce the number of suppliers
- **Interest**
  - Improve management of cash
  - Reduce inventory levels
  - Improve receivable process
  - Improve payable process
  - Minimize fixed asset investments
- **Taxes**
  - Optimize tax process to obtain tax advantages
  - Prevent overpayment of taxes
  - Optimize inventory levels

Many other projects show up when we look at our "hidden costs" within the business. These costs are a result of poor performing business processes. For example, flaws in new product development show up in delayed commercialization and delayed "kills" on new product projects, both of which lead to missing your growth goals. Your poor sales process will lead to lost orders and longer lead times. Your payables, receivables, and inventory process, if not efficient and effective, lead to cash flow issues.

We will be discussing in further detail the process of selecting the right projects to move you forward in your strategic actions. But once a good set of projects is identified and quantified financially, the next step is to deliver the money.

DELIVERING THE MONEY

Now comes the fun part—delivering the money. You have just spent a lot of time understanding your external realities, creating a strategy to take the offensive, and you've set some financial targets to reflect your strategic goals. You've then assessed your business processes and identified a set of important project clusters—areas that one or more projects will address.

Now, with the help of your Project Champions, your organization will identify specific projects that are linked to your financial targets and strategic goals. Each function in your organization will assign projects to Six Sigma Black Belts or Green Belts, and these project managers will be accountable for the completion of the projects. Your financial group will bless each project to ensure that financial impacts are realistic and can be achieved. Now the bottom-line focused Six Sigma starts. Your Black Belts and Green Belts will be given training in the roadmap and tools specific to the need of their projects.

For example, a marketing Black Belt will attend Marketing for Six Sigma training and learn the Marketing Six Sigma roadmap and tools. Your product development people will do the same with Design for Six Sigma (DFSS). Your manufacturing and transactional people will attend DMAIC (Define, Measure, Analyze, Improve, and Control) training in process improvement. Some of your organization will attend Lean Enterprise training or integrated Lean Sigma training.

In fact, there are over 20 different Six Sigma roadmaps that are available to address just about any strategic process-oriented problem you may have. Figure 2.6 depicts a typical business with three functions: to identify, provide, and communicate value. You will see seven Six Sigma macro roadmaps.

These roadmaps allow you to take Six Sigma into virtually every part of your company. The challenge of launching a Six Sigma program is to determine how to phase the seven roadmaps into your company while leveraging the right roadmaps on the right strategic goals.

An example of one roadmap, the DMAIC roadmap (sometimes known as the operations roadmap), consists of the following steps:

- **Define:** Identify the scope of the problem and estimated benefits of the solution.
- **Measure:** Measure the current variation of the performance data included in the problem.
- **Analyze:** Find potential sources of variation of the performance data.
- **Improve:** Verify, control, and optimize the sources of variation of the performance data.
- **Control:** Establish a system of controls to manage the gains of the solution.

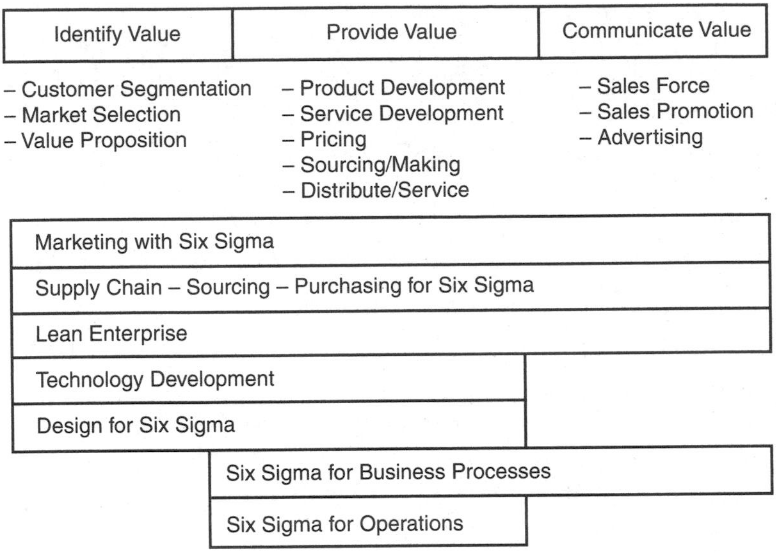

Identify Value	Provide Value	Communicate Value
– Customer Segmentation – Market Selection – Value Proposition	– Product Development – Service Development – Pricing – Sourcing/Making – Distribute/Service	– Sales Force – Sales Promotion – Advertising

Marketing with Six Sigma
Supply Chain – Sourcing – Purchasing for Six Sigma
Lean Enterprise
Technology Development
Design for Six Sigma
Six Sigma for Business Processes
Six Sigma for Operations

Figure 2.6 Six Sigma macro roadmaps to support the entire business.

So, each roadmap is a system made up of a series of discrete steps that lead to a solution to the problem. Each roadmap is set up to make decisions based on facts and data. Each roadmap drives toward delivering money to the bottom line.

In delivering the money, you can plan on training between 0.5% and 4% of your population as Black Belts and between 10% and 30% of your population as Green Belts. The percentages could easily be higher depending on the broadness of the scope of your Six Sigma deployment. Each Black Belt should deliver between $250K to $1.2 million per year and each Green Belt should deliver between $100K and $500K per year. Each completed project should have a financial signoff.

As you might have guessed, while Six Sigma is highly effective in delivering money, you can still expect a huge commitment of your own time and the time of your senior leaders to create the infrastructure and processes to support your new Six Sigma efforts. The remainder of this book will provide you with details into every milestone you must achieve to launch a Six Sigma program in 90 days (or less).

3 Six Sigma Launch Philosophy

With Ian Wedgwood

This chapter covers some general philosophical principles that have driven successful Six Sigma deployments in the past. Six Sigma is more than focusing on the money, finding the money, and delivering the money, as mentioned in the previous chapter. Six Sigma deployments, if done well, actually end up changing the culture of the organization. This was clear in the Motorola experience.

In just a couple of years, everyone in the company knew how to calculate a Sigma and was driven by the metric, defects per unit (dpu). Companies that have been successful with Six Sigma find that they operate much differently than before Six Sigma. For example, companies hear, "Where's the data to support that idea?" more often. The demand, "We've got to put some Black Belts and Green Belts on that issue," is heard frequently. Project identification and prioritization become more systematic and timely. Reward and recognition programs are ubiquitous, mainly because there is always some amazing result to be proud of. The nature of Six Sigma means you have to get better at a lot of things before Six Sigma is successful.

Six Sigma represents the methodology for pursuing breakthrough improvement in business performance. Figure 3.1 shows some actual data from a Six Sigma company (AlliedSignal). The graph depicts the company's 1995 earnings per share of publicly traded stock, and then compares 1995 results to its 1996 earning per share. This image, the financial bridge, shows the real difference from 1995 performance to 1996 performance directly from Six Sigma activities. AlliedSignal started an extensive Six Sigma initiative in 1994. This figure shows that AlliedSignal achieved about a 14 percent growth in earnings per share because of the Six Sigma activities during 1996 alone.

Six Sigma accounted for approximately $1.43 per share. Notice that the activities marked "Plant A at Full Capacity" accounted for $.03 per share. Not bad for a 23-year-old chemical engineer from Louisiana State University and her plant team. Her team optimized a new chemical plant in Baton Rouge after over a year of the "experts" trying to figure things out. These results were directly transferred to their sister plant in France, adding another $.05 cents per share.

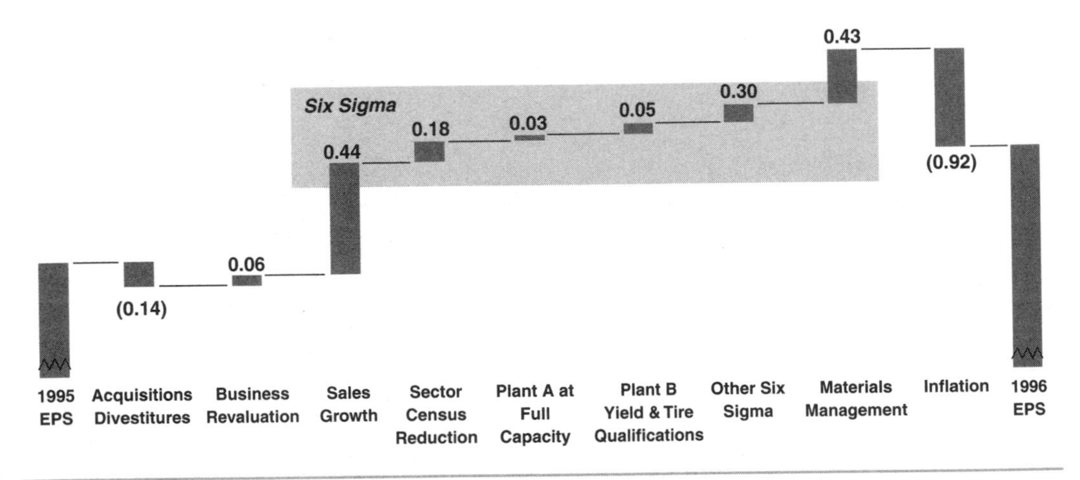

Figure 3.1 The Six Sigma financial bridge tracking the financial impact of Six Sigma on earnings per share from 1995 to 1996.

THE NATURE OF A SIX SIGMA DEPLOYMENT

Now that we've seen that Six Sigma is correlated to business and financial breakthroughs, there are many components that make up a Six Sigma deployment. The first step of the deployment philosophy is to establish the business. Just as any new business needs a business plan, so does deploying Six Sigma. Six Sigma—especially when including Lean Enterprise—includes many different business applications.

While Six Sigma addresses the accuracy of systems and Lean Enterprise addresses the speed of systems, the two methods integrated provide a powerful combination. At least three applications of Six Sigma combined with Lean Enterprise include (1) manufacturing operations, (2) business process support, and (3) new product development. Other specific applications include marketing, sales, supply chain management, safety, and enterprise systems. So, deciding the scope of the Six Sigma deployment and where to focus is critical at the onset of the deployment. Therefore, different processes use different forms of Six Sigma and their outcomes, as follows:

- Transactional Six Sigma for business processes: Optimizing process flow or designing new processes.
- Operational Six Sigma: Optimizing product and process flow.
- Lean Sigma: Products and special tools for process flow, speed, and accuracy.
- DFSS (Design for Six Sigma): Products, operations, and market windows.

In Figure 3.2, three typical core business processes (sales, new product development, and operations) are shown, along with the appropriate Six Sigma application. Here we see just a small fraction of the core business processes in a typical company—sales, product development, and operations. You notice the option to track your Six Sigma projects by determining to which process box the project applies. Figure 3.3 shows for each box in each process in Figure 3.2 a number in the lower box that indicates the number of Six Sigma projects identified for that particular process step. The figure shows this company is covering operations very well, but may change the focus to development and sales. If you have empty boxes, your Six Sigma program is not serving that part of the company.

So, now, when you line up your important core processes, you can easily see how you might overwhelm your Six Sigma program from the start. The secret is, based on the external realities, prioritizing the core processes and starting with a relatively few important core processes at the beginning allows you to evolve your Six Sigma program over a few years.

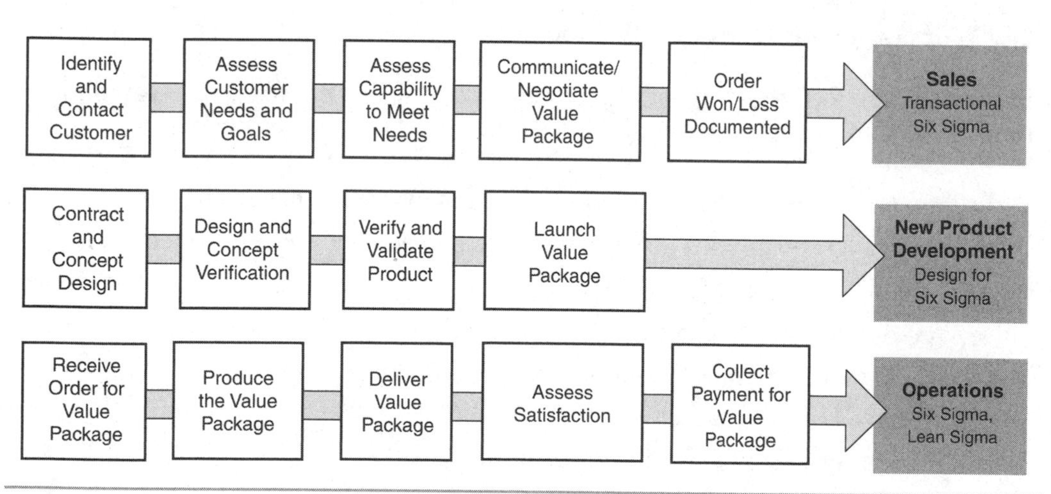

Figure 3.2 Matching Six Sigma form to the business core process.

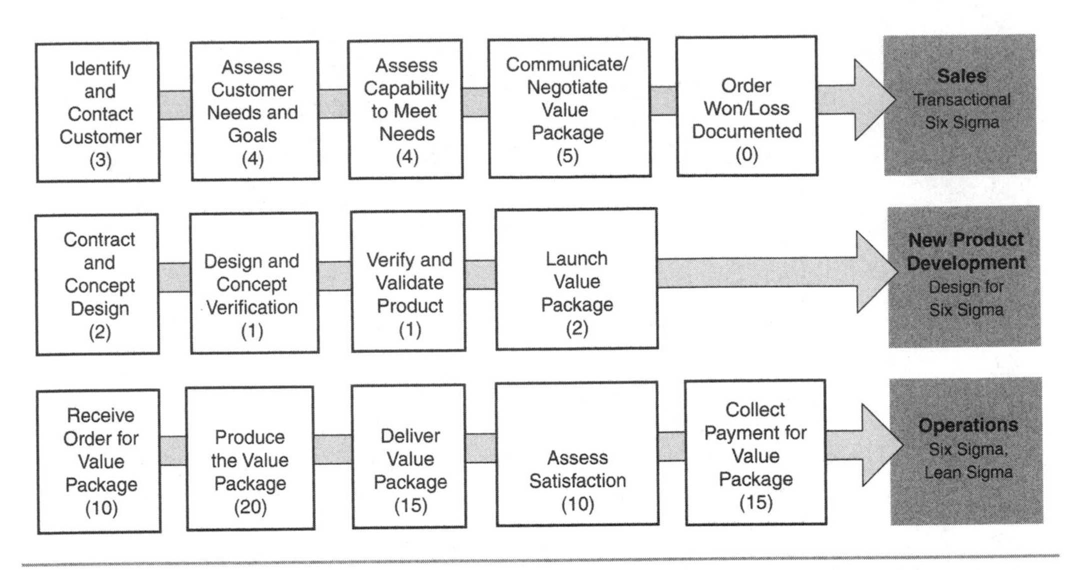

Figure 3.3 Matching Six Sigma to the business core process.The numbers in the lower center of each box indicates the number of Six Sigma projects that are underway.

For example, AlliedSignal was directed to focus the first year of Six Sigma (1994) on manufacturing operations. In 1995, Six Sigma was moved into R&D and product development. In 1996, transactional Six Sigma was initiated. AlliedSignal knew that if it tried to do too much too soon, it would not be successful.

SIX SIGMA DEPLOYMENT TIMING

Just to start an effective Six Sigma deployment, the company has to consider, at minimum, these activities:

- Selling the concept to the organization: Internal marketing plan.
- Systematically attain quick results (6–12 months) early in the deployment:
  - Lean and Kaizen activity.
  - Well-defined strategic projects.
- Create a structured project identification methodology:
  - Prioritized list of projects.
  - Define a "project hopper" for future projects.

    - Strategic mapping and assessment activities to focus the projects in the right areas.
- Launch a supporting infrastructure:
    - HR, finance, project tracking and reporting, Six Sigma steering teams, and Champions.

Of course, there are many additional milestones, but the preceding gives you an idea of the magnitude and the complexity of launching a successful Six Sigma deployment. The purpose of this book is to give you definitive guidelines with which to launch Six Sigma within 90 days. Now it's time to look at the philosophical architecture.

KOTTER'S PHILOSOPHICAL DEPLOYMENT ARCHITECTURE

I have drawn the Six Sigma deployment architecture from the fine work of Harvard Professor John P. Kotter and his book, titled *Leading Change*, published in 1996. The book is excellent and, because driving change is the most difficult task in a company, the book took the corporate world by storm. Kotter keeps things clear and reasonable. He was, in fact, invited to address the senior leadership of Larry Bossidy's AlliedSignal soon after Six Sigma was deployed. It is clear to me that Larry had used Kotter's ideas and produced terrific results.

John Kotter focused on the idea of transforming the organization. As a Harvard professor, in his studies, he'd seen common ways that organizations fail in their attempts to execute a transformation. He also defined eight stages that, if followed, will almost guarantee your transformation (i.e., Six Sigma deployment) will be successful. The following are Kotter's eight stages:

1. Establishing a Sense of Urgency.
2. Creating a Guiding Coalition.
3. Creating a Vision and Strategy.
4. Communicating the Change Vision.
5. Empowering Broad-Based Change.
6. Generating Short-Term Wins.
7. Consolidating Gains and Producing More Change.
8. Anchoring New Approaches in the Culture.

I will now address each of the preceding stages to launching a Six Sigma deployment.

KOTTER STAGE 1: ESTABLISHING A SENSE OF URGENCY

Establishing a sense of urgency is the toughest part of deploying Six Sigma. As Jim Collins says in his book, *Good to Great*, "Good is the enemy of great." Kotter says in a complacency-filled organization, change initiatives are dead on arrival. We'll take a quick look at the ever-present gremlin, complacency. Looking at complacency, several sources appear:

- Absence of a major critical crisis.
- Metrics that focus employees on narrow functional goals.
- Low overall performance standards.
- Internal measurement systems that focus on the wrong performance standards.
- Human nature, with its capacity for denial, especially if people are already busy or stressed.

To create a sense of urgency, some of these sources must be attacked directly. But, let's go back to the three-component business model, which links (1) external realities, (2) financial targets, and (3) internal activities.

Now there are at least two ways to create a strong sense of urgency. The first way is to identify a crisis. This crisis will be identified by clearly understanding the concept of external realities. Identifying a crisis creates emotional linkage to the change initiative. A crisis may be derived from at least two different sources:

1. The current organization is under siege now or will be soon.
2. The organization will not exist in the long run if things don't change.

Identifying potential opportunities for success provides the second way to create a strong sense of urgency. Convincing people that the change initiative, Six Sigma, will move the company to greatness is a realistic approach without having to dwell on or create a crisis. The final success of the launch is tied to significant financial targets of the program. That alone will generally provide the sense of urgency that is needed to launch a change initiative.

During AlliedSignal's 1994 launch of Six Sigma, there were significant financial goals established in productivity and growth. Fred Poses, then the president of the engineered materials sector of AlliedSignal (currently, Fred is the CEO of American Standard), used an optimum combination of opportunity and tough financial targets.

The opportunity, he declared, was based on some calculations literally handwritten on the back of an envelope while eating dinner in a restaurant. From his analysis, he estimated that the combination of lowering the sector's cost of poor quality and improving manufacturing capacity and yield would lead to at least a $1 billion savings over three years.

I know from my experience leading the Six Sigma effort for Fred that I had a specific goal of $150 million of pretax income by the end of the year. To reinforce that goal, I always heard the same question from Bossidy at the end of our one-on-one training sessions: "Steve, when will all this activity hit the bottom line?" I left the meeting with a real sense of urgency.

Creating a sense of urgency is important, but leading the Six Sigma deployment with courage is the real success factor. Both Larry Bossidy and Fred Poses went through a lot of training and knew the guts of Six Sigma, and more importantly believed in what Six Sigma could do for the company. They both invested a huge amount of time communicating the initiative. I personally watched Fred listen to each of his 30+ general managers explain to him in excruciating detail who their Black Belts were, what the projects were, and how much the projects would yield financially.

Larry held extensive quarterly Six Sigma reviews across AlliedSignal without fail. Larry spent at least 30 minutes with each class of Black Belts that I was training. Most importantly, both Larry and Fred picked the right people to drive the program. The Six Sigma Champions with whom I worked were all accomplished, respected, and passionate about Six Sigma. But it all worked because of a clear sense of urgency right at the beginning. Results backed up Larry and Fred's efforts. AlliedSignal achieved a two-year financial impact of Six Sigma amounting to about $880 million. Stock values rose accordingly. Six Sigma helped AlliedSignal become a premiere company.

Another example of creating a sense of urgency is 3M. Jim McNerney, a former GE senior executive, was named the new CEO. Within a couple of weeks of taking office, Jim had discovered 3M had been spending a lot of time and resources investigating Six Sigma. Jim made it easy for this group—he dictated that they would launch a three-day Six Sigma training session for the top 100 senior leaders of the company within about two weeks. Because Jim knew the guts of Six Sigma from deploying it into two divisions of GE, when I talked to Jim about my consulting company doing the leadership training, he said," Yeah, I broke some glass around here." Jim had helped his staff—which had worked on trying to decide whether or not to do Six Sigma for over a year—to decide to do Six Sigma in only a few days.

He also explained to me that he wanted the fastest Six Sigma deployment in history. And he did it! Within three weeks of the leadership session, which got a standing ovation, hundreds of projects and Black Belts were launched. Now, 3M literally has thousands of projects going simultaneously, and 3M's performance in the stock market matches the effort. Jim was able to be so courageous because he knew in his heart that Six Sigma was the right thing for 3M.

Jim set tough financial expectations for the Six Sigma deployment: (1) getting products to market faster; (2) improving cash flow; and (3) improving productivity. He grew Six Sigma into an additional initiative—3M Acceleration—by meshing Design for Six Sigma into 3M New Product Introduction (NPI) process. This was a great move since this initiative built on 3M's tradition of product innovation.

One other way to establish a sense of urgency is to be clear about the cost of procrastinating in Six Sigma. In an early presentation to the leadership of AlliedSignal's engineered materials sector, Fred Poses (President) and Donnee Ramelli (VP of Quality) presented the financial target, which was pretax income of $150 million additional dollars. The table looked like this:

Our Goal in 1995

1994 Income	$330M
1995 Income	$420M
Increase in Net Due to SS	$ 90M
Pretax Income for SS	$150M

The preceding chart said that, through effective Six Sigma deployment, the financial target was $150 million of pretax income—not a small number for a $4 billion sector. But then the urgency was actuated. Fred and Donnee presented a table that expressed the average monthly benefit that Six Sigma would have to deliver, depending on when each organization started. This was presented at a meeting held in late 1994 to drive for results in 1995.

1995 Monthly Improvement Required

If We Start In	
January 1995	$12.5M/Mo.
April 1995	$16.7M/Mo.
July 1995	$25.0M/Mo.
October 1995	$50.0M/Mo.

The point was clearly made. If you sprint right at the beginning of training Black Belts, you had to average about $12.5 million a month to meet the $150 million goal. But if you waited until July to start, you'd have to average about $25.0 million per month, almost twice as much as the January implementers. This was very effective in getting the organization fired up to get started. In fact, even when we started two classes (or waves) of Black Belts in January 1995, I received several complaints from leaders who were asked to wait until the second class started to send their Black Belt candidates. They claimed the extra two weeks of waiting was going to kill them.

The sense of urgency is primarily the product of great leadership. One CEO claimed that when launching an initiative, "You have to change the people or change the people." For an initiative to be truly internalized, you must have all your senior leaders on board and at least the vast majority of everyone else. The CEO had replaced a number of his senior leaders at AlliedSignal to communicate the sense of urgency. Fred Poses replaced one plant manager very quickly because that manager didn't take Six Sigma seriously. Believe me, all the other plant managers became extremely interested in the initiative. The following chapters address in detail establishing a sense of urgency:

- Chapter 5, "Strategy: The Alignment of External Realities, Setting Measurable Goals, and Internal Actions"
- Chapter 6, "Defining the Six Sigma Program Expectations and Metrics"

KOTTER STAGE 2: CREATING A GUIDING COALITION

Creating a guiding coalition leads to establishing a group of the right people who have enough power and leadership ability to lead the Six Sigma deployment. There is also the challenge of creating a team of people who have the capability of working as a team. Companies successful in deploying Six Sigma have a clear commonality. They are smart enough to put a strong and credible leader at the helm of the deployment.

In AlliedSignal, Larry placed a newly acquired Vice President of Manufacturing, Richard Schroeder, in charge of the deployment. Rich was very experienced in Six Sigma from his history with Motorola, and he also drove Six Sigma into ABB as well. Rich had a double reporting structure: (1) CEO and (2) VP of Quality—Jim Sierk. So the combination of starting with someone who knew the guts of the program and giving him a clearly high reporting line set the stage for success.

The next step was to create several "boards of directors" or steering teams in each of the three AlliedSignal sectors: (1) aerospace; (2) automotive; and (3) engineered materials. The president of each sector was requested to identify Six Sigma Champions. This group of about 60 people attended a one-week training session very early in the deployment. Because the original focus of Six Sigma in AlliedSignal was on manufacturing operations, these Champions were the cream of the manufacturing crop. They were all leaders with credibility and respect in their sectors.

Some other examples of starting off Six Sigma with a guiding coalition would include Cummins, 3M, and Celanese. Each of these companies carefully selected the Six Sigma leader. The Cummins (a manufacturer of diesel engines) CEO, Tim Solso, selected Frank McDonald to lead the program. Frank was probably Tim's best operations expert, so Tim knew Frank would make it happen. Frank, of course, was a little confused at first, being moved from the operations role he loved, to being the corporate quality guru. He initially thought he was being punished, but as he learned the guts of the program, he soon realized he could change Cummins into a different company, which he did.

At 3M, Jim McNerney selected Brad Sauer, a highly respected up-and-coming business leader. Brad was a strong leader and the perfect person to lead the program. Brad, after leading the first two years of Six Sigma, was given a business to lead. At Celanese, a large chemical manufacturing company, the President, David Weidman, was a former AlliedSignal business leader working for Fred Poses.

Dave knew operations, the guts of Six Sigma, and knew in his heart that Six Sigma was the right thing to do strategically for Celanese. He put his best operations person, Jim Alder, at the helm of Six Sigma. Once again, Jim, being an ops guy at heart, thought he was being punished with his new role. It was not long before Jim figured out that, by driving Six Sigma, he could make Celanese a better company, which he did.

Carefully selecting the leader of the initiative cannot be underestimated. The selection of the Initiative Champion is the first shot across the bow in a Six Sigma deployment. The person selected will tell the entire organization how important the initiative is. Larry Bossidy, in his recent book, *Confronting Reality*, says that picking the right people to implement the initiative is extremely important. Look to his book to get his insights on this matter.

In the three companies discussed, the CEOs selected the right leader carefully, and then the new leaders surrounded themselves with an effective deployment team—usually referred to as Deployment Champions. In all cases, the deployment team received at least a week of Six Sigma training and participated in the deployment planning—generally with the help of an outside consulting group. These CEOs also changed the reward structure, usually putting a hefty percentage of their bonuses on results from deploying Six Sigma.

There is also the steering team side of deployments. While at AlliedSignal, I created several steering teams that were accountable for some aspect of Six Sigma. These steering teams literally served as my boards of directors for their part of the program. I found that if I created the steering teams correctly, I could effectively transfer accountability for results to the team. Some examples of steering teams were the following:

- Technology Steering Team
- Manufacturing Council
- Equipment Reliability Council
- Chemical Laboratory Council
- Financial Council
- Human Support Steering Team

These teams help define the appropriate Six Sigma roadmaps, devise the deployment plan for their Six Sigma program, determine the strategic direction of their program, and report results on financial targets.

The following chapters cover aspects of creating a guiding coalition in great detail:

- Chapter 8, "Defining the Six Sigma Infrastructure"
- Chapter 10, "Creating Six Sigma Executive and Leadership Workshops"
- Chapter 11, "Selecting and Training the Right People"
- Chapter 13, "Creating the Human Resources Alignment"

So, it's on to Kotter's third stage of creating major change.

KOTTER STAGE 3: CREATING A VISION AND STRATEGY

When you hear the word "vision," you probably immediately remember very long vision development meetings and, like most people, you hated it. I won't talk about how to create a vision, even though I have had good luck with vision meetings that lasted about 1.5 hours and produced a vision everyone in the room liked. But, when launching a major initiative like Six Sigma, the organizational leadership must have a clear and simple vision about what Six Sigma will do for the company.

I call these "elevator speeches." You meet someone on the elevator in corporate headquarters and they ask, "What's all this about Six Sigma?" If you can't answer that question before you hit the exit point, you don't have a clear idea of what Six Sigma is all about for your company. This is the vision presented by Fred Poses to his leadership for the engineered materials sector, which was based on AlliedSignal's corporate vision:

Performance Drives Results

A Premier Company Achieves Performance Every Year	
• Sales Growth	12+%
• Earnings Growth	20+%
• #1 with Customers and Employees	
• Operational Excellence (Six Sigma)	

This vision was simple, clear, and effective. A leader could easily explain it on the elevator. For Paul Norris, then CEO of WR Grace, the vision was easy. He put an overhead slide on the projector at the leadership training session, and it had one number on it—$100 million—and that's what he said he expected from the program. Because Paul was an alumnus from the AlliedSignal Six Sigma experience, everyone in the room knew he was serious and knew what he was talking about.

The strategy part of launching Six Sigma is an output of linking the three business model components: (1) external realities; (2) financial targets; and (3) internal activities. The strategy is made up of the initial Six Sigma deployment plan, the long-term strategic development of Six Sigma, and developing the roadmap for leading Six Sigma over the long term. The following chapters address creating a vision and strategy:

- Chapter 4, "Getting Early Support: Selecting a Six Sigma Provider"
- Chapter 5, "Strategy: The Alignment of External Realities, Setting Measurable Goals, and Internal Actions"
- Chapter 6, "Defining the Six Sigma Program Expectations and Metrics"
- Chapter 7, "Defining the Six Sigma Project Scope"
- Chapter 8, "Defining the Six Sigma Infrastructure"
- Chapter 9, "Committing to Project Selection, Prioritization, and Chartering"
- Chapter 13, "Creating the Human Resources Alignment"
- Chapter 15, "Leading Six Sigma for the Long Term"

KOTTER STAGE 4: COMMUNICATING THE CHANGE VISION

Communicating the upcoming change widely and clearly is absolutely necessary for change to occur. Communication plans are complex because they have to use a wide variety of communication forums. But there are some guidelines on effective communication of the Six Sigma initiative.

The first guideline is simplicity. Making the change clear and exciting is best with simple communications. Repetition of the simple message is crucial. Larry Bossidy and Fred Poses were the best I've seen. In every meeting and every event they attended, they had a short, dynamic message. They both relied on multiple forums and media. Fred was very consistent about calling and leading leadership conferences addressing Six Sigma.

Both Larry and Fred attended multiple Black Belt and Green Belt sessions to deliver their message. Larry once said you have to communicate an initiative so often that you finally want to throw up every time you say it. Well, I assume he exaggerated, because I never saw him throw up. The point was understandable—you might be willing to commit to the change if you're always talking about it in a repeating message.

And, finally, there must be many give-and-take opportunities for all members of the guiding coalition. And, Kotter recommends the creation of a verbal picture to use often. You will see the following chapters address communicating the vision:

- Chapter 10, "Creating Six Sigma Executive and Leadership Workshops"
- Chapter 12, "Communicating the Six Sigma Program Expectations and Metrics"
- Chapter 13, "Creating the Human Resources Alignment"
- Chapter 14, "Defining the Software Infrastructure: Tracking the Program and Projects"

The following is an example of a vision for Six Sigma actually developed at the onset of the Six Sigma deployment:

Our Six Sigma Commitment to Drive
Growth
• Understanding the voice of the customer
• Value proposition
• Faster technology development
• Faster product commercialization
Cost/Productivity
• Quality improvements
• Cost of poor quality improvement
• Capacity improvement (without capital)
Cash/Working Capital
• Payables
• Receivables
• Inventory

KOTTER STAGE 5: EMPOWERING BROAD-BASED CHANGE

You're in luck here, for Six Sigma capitalizes on empowering employees to support the company's goals. That's the only way your Six Sigma program is going to work. Let's go back to the three-component business model, which links (1) external realities; (2) financial targets; and (3) internal activities. Simply, by understanding your external realities and creating a strategy to address those realities and setting your financial targets to reflect your external realities, you must align your internal activities to get the results you need.

After your strategy is defined to address the challenges of your external reality, your guiding coalition will work on identifying and prioritizing projects that directly support the strategy. Of course, the guiding coalition will need a large amount of training to make this happen.

After these projects are identified, each organization must select and train the right people to apply Six Sigma to complete those projects. The major focus is naming and training Black, Green, and Yellow Belts. Great projects drive the training success, which consists of defining great projects and resources for the project teams before the student attends his or her first training session.

When the Belts attend training with projects that are obviously tied to business and strategic success, the training becomes a way to accomplish those goals. The challenge of the first 90 days is planning for the Black, Green, and Yellow Belts in Operations, R&D, and Transactional areas.

Now, when you consider that AlliedSignal trained over 450 Black Belts during the first full year of Six Sigma, that means there were at least 450 resourced projects with an average of 10 members on each project team. Thus, at least 4,500 people were empowered for broad-based action. That didn't include Green Belts because not many were trained the first year. Every member of every team understood where the project fit in helping to make AlliedSignal a premier company.

So, after you understand what your company's performance gaps are based on the external realities, you define a strategy to close those gaps, select projects identified in fixing the gaps, and train Belts and apply resources to their teams; you automatically empower employees for broad-based action. That's when Six Sigma gets very exciting. The chapters addressing employee empowerment are as follows:

- Chapter 5, "Strategy: The Alignment of External Realities, Setting Measurable Goals, and Internal Actions"
- Chapter 6, "Defining the Six Sigma Program Expectations and Metrics"
- Chapter 7, "Defining the Six Sigma Project Scope"
- Chapter 8, "Defining the Six Sigma Infrastructure"

- Chapter 9, "Committing to Project Selection, Prioritization, and Chartering"
- Chapter 11, "Selecting and Training the Right People"
- Chapter 13, "Creating the Human Resources Alignment"
- Chapter 15, "Leading Six Sigma for the Long Term"
- Chapter 16, "Reinvigorating Your Six Sigma Program"

KOTTER STAGE 6: GENERATING SHORT-TERM WINS

This phase consists of planning for, creating, and rewarding those early wins. In reality, early results are the best way to drive change. However, the results must be clear and convincingly substantial. For example, within the engineered materials sector of AlliedSignal, the six-month yield of Six Sigma projects was about $50 million. Six Sigma definitely had the sector's attention. Every plant manager was almost evangelistic about the program.

There is clearly an opportunity to launch your first wave of Six Sigma training within 90 days of your first deployment milestone and start posting financial results within another 3–6 months. The deployment strategies documented in this book has short-term wins as its focus. The chapters that apply to generating short-term gains are as follows:

- Chapter 4, "Getting Early Support: Selecting a Six Sigma Provider"
- Chapter 6, "Defining the Six Sigma Program Expectations and Metrics"
- Chapter 9, "Committing to Project Selection, Prioritization, and Chartering"
- Chapter 11, "Selecting and Training the Right People"
- Chapter 13, "Creating the Human Resources Alignment"
- Chapter 14, "Defining the Software Infrastructure: Tracking the Program and Projects"

KOTTER STAGE 7: CONSOLIDATING GAINS AND PRODUCING MORE CHANGE

Pending a successful Stage 6 that produces dramatic early results, the next step is to grow the Six Sigma initiative with other focused projects. For example, AlliedSignal focused on manufacturing operations during the first year, and then added emphasis to growth by launching Design for Six Sigma (DFSS), and later started Transactional Six Sigma. Each new project has the same accountability to the financial targets as the first project.

This stage migrates Six Sigma to include all parts of the company. Usually, the first migration for U.S.-based companies is to launch Six Sigma in Europe and Asia. As Kotter quotes, "Culture changes only after you have successfully altered people's actions, after the new behavior produces some group benefit for a period of time, and after people see the connection between the new actions and the performance improvement."

People will see over and over again Black Belts dropping $250,000 to $1,000,000 to the bottom line—project after project. They will either want to be on a project team or be trained as a Black Belt. Simply, however, as stock prices continue to go up, change becomes easier. The chapters addressing consolidating gains and producing more change are as follows:

- Chapter 5, "Strategy: The Alignment of External Realities, Setting Measurable Goals, and Internal Actions"
- Chapter 7, "Defining the Six Sigma Project Scope"
- Chapter 8, "Defining the Six Sigma Infrastructure"
- Chapter 12, "Communicating the Six Sigma Program Expectations and Metrics"
- Chapter 14, "Defining the Software Infrastructure: Tracking the Program and Projects"
- Chapter 15, "Leading Six Sigma for the Long Term"
- Chapter 16, "Reinvigorating Your Six Sigma Program"

KOTTER STAGE 8: ANCHORING NEW APPROACHES IN THE CULTURE

After the Six Sigma deployment is successful, you have to start thinking about what you need to do to make sure the company is doing Six Sigma 10 years from now. Even the Six Sigma pioneer, Motorola, which had launched Six Sigma in 1978, found Six Sigma lacking in the late 90s. This phase leads to the understanding that the change initiative—in this case, Six Sigma—has to be reviewed at least every three years with respect to reinvigorating the initiative.

Anchoring an initiative is directly related to the infrastructure and systems developed to support the initiative. Fortunately, the very nature of Six Sigma addresses developing streamlined processes and systems. So the systems developed to support Six Sigma should be user friendly and effective.

By consistently pointing out the relationship between Six Sigma activities and organizational performance, a connection is made between the new actions equating to closing competitive gaps. A key to anchoring Six Sigma into your culture will be to develop future leaders who have a realistic view of the value of Six Sigma. It turns out that Six Sigma is a great leadership development program. Jack Welch, while CEO of GE, expected that every high-potential leader would be certified in Six Sigma. Most of GE's future leaders know firsthand the value of the program and will not let it die.

Six Sigma must have something going for it, judging by the number of CEOs who left Six Sigma companies for a non-Six Sigma company in order to introduce Six Sigma. This list is long, which indicates that Six Sigma should not be difficult to anchor in the culture, but you'd better pay attention to doing that. Chapters addressing anchoring new approaches in the culture are as follows:

- Chapter 5, "Strategy: The Alignment of External Realities, Setting Measurable Goals, and Internal Actions"
- Chapter 6, "Defining the Six Sigma Program Expectations and Metrics"
- Chapter 8, "Defining the Six Sigma Infrastructure"
- Chapter 12, "Communicating the Six Sigma Program Expectations and Metrics"
- Chapter 13, "Creating the Human Resources Alignment"
- Chapter 14, "Defining the Software Infrastructure: Tracking the Program and Projects"
- Chapter 15, "Leading Six Sigma for the Long Term"
- Chapter 16, "Reinvigorating Your Six Sigma Program"

SUMMARY

We've covered Kotter's model for leading change. The model clearly works for not only Six Sigma, but any initiative. You should always review Kotter's eight stages and do them sequentially. After you succeed at deploying Six Sigma, you will then have a whole organization that is equipped to launch major change initiatives into the company. You will not only have a viable Six Sigma program upon which you are getting a huge return on investment, but you will have created a core competency to effectively launch change initiatives.

4 Getting Early Support: Selecting a Six Sigma Provider

Here is the basic corollary in deploying Six Sigma. It's probably a good idea to bring in a Six Sigma provider (consulting group) to add to your resources. I have seen few large companies deploy Six Sigma successfully when using only internal resources. Would you rather have a few internal people who have seen one or two Six Sigma deployments, or hire an external group with experience with a few dozen Six Sigma deployments? The consultants become variable cost rather than fixed cost. Even 3M, which had great internal resources, chose to use external consultants to kick-start their program. Because of 3M's internal resources, they were able to internalize the program quickly.

Of course, we've all had bad experiences with consultants—usually from the Big Six consulting firms. The standing joke is that a management consultant comes in, asks for your watch, looks at the watch, and tells you what time it is. Consultants have also been accused of having an interesting comparison to sea gulls.

But the fact of the matter is that just about all major Fortune 500 company deployments have used an outside provider. At a minimum, they'll hire in outside resources from successful Six Sigma companies to buy the capability. And those resources will usually have great depth in Six Sigma deployment experience, perhaps having served as a consultant at one time or another. These new resources will probably have a limited experience in actually deploying Six Sigma. Even Motorola used some outside consultants to put together the original Six Sigma and statistical training material necessary to deploy skills throughout the company. Companies that used outside consultants include AlliedSignal, Honeywell, GE, Dow, DuPont, Ford, 3M, Nokia, Sony, Johnson and Johnson, and many others. Because of the large numbers of Six Sigma deployments, a select few Six Sigma consulting companies have a wide range of experiences in working with large companies to deploy Six Sigma.

My company noticed an interesting slogan displayed by a local plumbing company on the side of one of its trucks, "Often out promised, and seldom out delivered." That became our slogan as well. You'll find that Six Sigma providers will tend to say they can do anything you ask for. The challenge is to cut through all the bull and get to the truth. Unfortunately, there are at least two external consulting ghosts in business that haunt Six Sigma deployments: consulting associated with TQM (Total Quality Management) and business process reengineering. Most of us remember those two change initiatives and shudder about going through those again. TQM was one of the worst, but TQM is also an initiative to which Six Sigma is often compared.

These TQM consulting engagements usually lasted for years and included training events everywhere in the company. One of American business's fatal management flaws is to confuse activity with results. During a TQM deployment, there was always a huge amount of activity: training sessions, reward and recognition sessions, TQM conferences with thousands of people involved. The problem was that most TQM companies rarely saw better business performance.

Generally, the expectations for business performance improvement were not clearly stated and not well tracked. The TQM team typically used the number of people trained as the driving metric. Training doesn't have an impact on the organization unless training drives new organizational behavior in consistent and positive ways. For example, many companies trained their people in Stephen R. Covey's *Seven Habits of Highly Effective People*. Those training programs fell flat because the leaders of the organization were not clear about the behavior changes expected from the program.

Larry Bossidy called TQM from his experience, "Mugs and Hugs." His point was that there was a whole lot of action, but no impact on the bottom line. And Larry was right. A CEO for a large carpet manufacturer told me that when he added up all the alleged saving from his TQ teams together, the number would sometimes dwarf the actual revenue for the year. This is a classic symptom of the business leadership not linking the program to actual financial targets.

Arthur D. Little surveyed 500 manufacturing and services companies and found that only a third felt TQM was making a significant impact on their businesses. Likewise, A.T. Kearney studied 100 British firms and only a fifth felt TQM left a tangible legacy of business improvement. And I've seen other surveys that reported even worse results than that. TQM was weak because its activities were never linked to the company's external realities, strategies, or to clear financial targets. One of our client companies, a large railroad company, was wrestling with deploying Six Sigma after their relatively unsuccessful launch of TQM. They finally created the slogan, "Six Sigma will put brains into TQ."

Both Larry Bossidy and his mentor, Jack Welch, consciously and publicly shied away from TQM. Interestingly, they both got right on board with Six Sigma, and both admit it was worth it. In fact, in Jack's book, *Winning*, Jack says, "These days, with Six Sigma being increasingly adopted by companies around the world, you can't afford not to understand it, let alone not practice it."

Then, in the 1990s, reengineering replaced TQM as the next business panacea. While the idea of reengineering made sense, the methodology was either nonexistent or weak. Reengineering was the method to streamline business processes and eliminate waste. However, it was not related to external realities or financial targets.

In a *Harvard Business Review* article, the authors studied over 100 companies doing reengineering. These companies claimed some improvement in business processes but little impact on overall business performance. In fact, reengineering developed a reputation as a method to adjust an organization's headcount to a lower, more cost-effective level. Again, reengineering was not effectively linked to the external realities, financial targets, or strategy.

Michael Hammer, developer of reengineering, claims Six Sigma is not as effective, and, yet, every company that has deployed Six Sigma will claim significant business results. In fact, Hammer's "official definition" of reengineering from his book, *The Reengineering Revolution*, is as follows:

> The fundamental rethinking and *radical redesign* of business *processes* to bring about *dramatic* improvements in performance.

This revolutionary definition easily describes Six Sigma. Hammer should have been ecstatic that Six Sigma provided the "how" to Hammer's "what to do." Since Hammer's 1993 book, *Reengineering the Corporation*, came out about the time AlliedSignal launched Six Sigma, in the second millennium Six Sigma is thriving. You don't hear about companies reengineering anything. In fact, the company's old reengineering group is usually right there to help with the Six Sigma deployment. Another consultant-heavy program bites the dust!

Six Sigma, Consultants, and Consulting Costs

So, why do consultants and Six Sigma work so well together? For one, Six Sigma is very heavy technically. Our Master Black Belts achieve the equivalent of a master's degree in advanced statistics and change leadership. Also, the corporate deployment protocols are well documented and accepted. Many of the Six Sigma consultants learned Six Sigma as a corporate employee involved in a Six Sigma deployment.

When investigating the ROI of a Six Sigma deployment, the consulting fees are only a small percentage of the improvement in your financial targets. For example, within two years after deploying Six Sigma, both AlliedSignal and GE earned about $880 million to the bottom line. Consulting fees were just a very small percentage of those gains. Richard Schroeder, Six Sigma leader at AlliedSignal, found that AlliedSignal easily paid for all the consulting costs for the first two years within the first six months of deployment. The ROI for consulting services is very quick, especially if you write the contract for consulting services correctly.

Intellectual Property. Six Sigma deployments consist of a huge amount of intellectual property (IP) because training is so crucial. My company alone has over 160 course offerings and over 1,000 course modules already developed over a period of eight years. Few companies have adequate resources with the capability to generate the IP, and generating IP internally would cost a lot of money and, more critically, a lot of time.

The training material must represent the best in instructional technology, so piloting mediocre training material is not a wise strategy. To engage a consulting group that has an IP portfolio, licensing the IP for use inside the company is faster, more effective, and more efficient when compared to developing the IP internally by internal resources. Plus, a good consulting company will accelerate the speed of internalizing the program by training your trainers.

As business leaders, we frequently grossly underestimate the time and actual cost of developing training materials and training support. We often overlook the idea that we don't have many people within the organization that have the core competency in Six Sigma needed to develop a comprehensive program. Besides, you are probably looking at a change initiative like Six Sigma because you really don't have the capability to drive that change residing within the company.

Some companies I've worked with have had a solid core of people who would quickly learn the guts of Six Sigma. 3M, for example, had a large number of statisticians from which to draw, and that group was converted to their primary Six Sigma training arm. But 3M still called on my company to provide the initial IP and initial training experiences, and 3M turned the program over to internal groups about 12 months later.

While 3M did this, my company still worked closely with 3M's internal talent to customize the training material to match 3M's previous good work in process improvement. 3M ended up with a state-of-the-art program very quickly and, as it turned out, very cost effectively.

Accountability for Results. The key to success in using a Six Sigma provider for your Six Sigma deployment is to achieve specific financial targets by working together toward the priority. I remember initially consulting with AlliedSignal. Fred Poses, President of the Engineered Materials Sector, told me that I would have an AlliedSignal office, business card, and phone number. I would also have AlliedSignal expectations.

He was clear in stating his expectation that he wanted his people to think I was one of them and I should work with them as if I was. He ultimately made me an offer to join AlliedSignal to direct the Six Sigma effort, and I joined AlliedSignal as an employee. This view of true partnership is the way the corporation and consultant relationship should be viewed.

CREATING A PARTNERSHIP

The process of creating a partnership with the Six Sigma provider is essential to the success of the deployment, both legally and in spirit. The provider must understand that they are accountable for results and maybe even rewarded for outstanding results. In a sense, the provider should feel like they work for and report to the company. The account manager should feel like and behave as one of your organization's senior leaders.

Transfer of Technology. The partnership represents the expectation to leave the company with the skills and capabilities to drive Six Sigma without the need of external consultants. This transfer of technology must happen quickly. Great care must be taken not to end up with a provider that establishes a beachhead into your company and stays inside for years. Two or three years should be the maximum expectation of the length of the Six Sigma engagement. Some companies have internalized Six Sigma in less than one year.

Steering Teams. Careful definition and use of steering teams and Champions to direct the provider are called for. Your own employees are, in fact, ultimately accountable for the success of the initiative, not the consultants. The provider should have a group representing your company functioning as a Six Sigma board of directors to which it reports.

Customization teams consisting of company personnel and provider personnel are common. The last thing you need is the perception that the consultants are running the Six Sigma deployment. The company needs to see clearly that their leadership is in charge and excited about it. Every activity should be perceived as close collaboration between the company and the providers.

The consultants will always be matched with internal resources with the consultants providing the initial drafts of the deployment plans. At all times, your internal resources must be aware that they have the accountability for both the success of the program and hitting the financial targets.

Program Reviews. You also should include scheduled program reviews to the provider's contract to ensure that corporate strategy and corporate financial goals are met. And it's reasonable to require the provider to do a customer survey about six months into the deployment and to share the results. Be sure to provide a list of people to be surveyed. If you want to know where the provider is weak and strong, these surveys tell it all.

If you are ever called as a reference for the provider, you'll want to be able to say that this provider was effective. Their costs were in line with the market, the provider saved you months in deployment time, the consultants knew what they were doing, and they helped beat financial targets by 20%. If you can say those things, you had a great provider.

IDENTIFYING, PRIORITIZING, AND SELECTING PROVIDERS

Identifying Providers. Identifying potential providers is uncomplicated in the age of the Internet. You simply search some keywords. The Six Sigma keywords will give you more providers than you know what to do with. In 1998, there were probably only five or six consulting companies specializing in Six Sigma deployments. Now there are hundreds. But determining which provider is the best for you is not so simple because there is a broad differentiation among these competitors. Some are excellent, and some are horrible. Web sites like Isixsigma.com should work well in your initial search activities.

Probably the best way to arrive at a short list is to poll the companies that have already deployed Six Sigma and find out from them who they used. It's especially helpful if you poll your own people to find out who they know and where. Personal friends or acquaintances can give you more accurate appraisals of their providers.

Other sources of information about providers are the numerous Six Sigma conferences, such as the ones sponsored by IQPC and ISSSP. These conferences are widely attended by consulting companies, and you can pick up materials and get some good face-to-face contact.

I provide a list of questions to ask at the end of this chapter. These providers will generally make formal presentations throughout the conference, so you are in the position to evaluate their technical capabilities.

The RFP: Prioritize and Select Providers. The best tool for making your way through this confusing market is to develop a Request for Proposal (RFP). The RFP will ensure consistency in the way you evaluate potential providers. It will also help you clarify what you need to create a successful Six Sigma program before you actually get started. I have included an example of an RFP format in Appendix A, "RFP Sample Format." You will probably find that you will use a combination of these to construct your final RFP.

RFPs are tricky. The purpose of the RFP is to get critical information about each provider in a systematic way and develop a short list of consulting companies to invite in for formal face-to-face presentations. If you ask for too much information, the prioritization process and selection process get cumbersome. If you don't ask for enough information, you may make the wrong decision.

Selecting Providers: The RFP Decision Matrix. Ultimately, being able to convert the provider responses on the RFP to scores on a decision matrix is valuable to the selection success. Table 4.1 shows an example of a decision matrix. The secret is to develop a matrix that follows your RFP outline. Notice the RFP elements are listed in the Criterion column. Your selection team will then weight the importance of each RFP criterion on a one-to-five scale and designate the weights in the Assigned Weight column.

Finally, each provider on the finalist list is rated on how well they address each criterion, again on a one-to-five point scale. Then the criterion weights and the provider weights are cross-multiplied and the products are summed. The provider with the highest total score is the frontrunner.

Table 4.1 Example of a Decision Matrix Comparing Three Potential Six Sigma Providers to Critical Criterion for Your Company

Criterion	Assigned Weight (1–5)	Service Provider Ratings (1–5)			
		1	2	3	4
Executive credibility	5	5	5	5	
Corporate deployment capability	4	5	5	3	
Combined years of Six Sigma experience	3	2	4	2	
Financial and documented results	4	4	4	2	
References	5	5	4	3	
Training portfolio	4	3	4	3	
Consulting capability	3	2	4	2	
Cost	3	1	4	2	
Knowledge of corporate culture	2	1	3	4	
Strategy linkage	4	3	2	3	
Project management capability	3	4	4	2	
Foreign language support	4	5	4	4	
Roadmap validity	3	2	5	3	

(continues)

Table 4.1 Example of a Decision Matrix Comparing Three Potential Six Sigma Providers to Critical Criterion for Your Company (Continued)

Criterion	Assigned Weight (1–5)	Service Provider Ratings (1–5)			
		1	2	3	4
Field and classroom support	5	3	5	2	
Course evaluation system	3	5	4	2	
Cycles of learning	4	3	4	1	
Material development	5	2	3	3	
Leadership training (material/flexibility)	4	3	2	4	
Deployment methodology	5	4	4	5	
License to material	3	3	4	3	
Assessment capability	2	2	5	4	
Total score		**262**	**308**	**234**	

For example, Table 4.1 shows an example of evaluating three potential Six Sigma providers. This particular company chose the following four criteria as the most important: Executive Credibility, References, Material Development, and Deployment Methodology. These were all rated as fives on the one-to-five scale. The numbers listed for the total scores (262, 308, 234) are calculated by adding the cross-multiplication of the provider score for each criterion and the criterion weight. This result indicates Provider 2 is the top provider in addressing the 21 criteria.

You have the option to make the decision matrix even simpler. Table 4.2 shows an example of a decision matrix that perfectly follows the actual outline of the RFP. This matrix ensures that everyone on the selection team reads each RFP carefully and makes the scoring process simpler. You might use the simple matrix for your initial screening and the more complex matrix for the final presentations. You can complete the complex matrix as each provider presents.

Table 4.2 Example of a Decision Matrix Comparing Four Potential Six Sigma Providers to Critical Elements of Your RFP

Criterion	**Assigned Weight (1–5)**	**Service Provider Ratings (1–5)**			
		1	**2**	**3**	**4**
Executive Summary					
Business Operations:					
• Facilities					
• Capabilities					
• Experience					
• References					
Materials:					
• Customization					
• Foreign Languages					
• License					
Training:					
• Portfolio					
• Coaching					
Experience					
Consultants					
Account Management Method					
Risks					
Total score		0	0	0	0

DIFFERENTIATING PROVIDERS

Differentiating providers can be a problem because there are so many Six Sigma providers in existence. The next few sections will provide some criteria for consideration when you are going through the throes of selecting a Six Sigma partner.

Corporate History. Asking the potential provider for a summary of their corporate history is insightful. How long have they been in business and what have they accomplished throughout their corporate life? Questions referring to the size of their corporation, the number of years they've been focusing on Six Sigma, the number of clients they've supported through corporate deployments, and their deployment philosophy should be addressed. The selection criteria are: are these guys real, are they a mom-and-pop shop or a real corporation, are they new to Six Sigma consulting, or do they have authentic expertise? Do they have a corporate headquarters or do they work out of garage?

Corporate Deployment History. Does the provider have a significant number of corporate deployments on record? Some Six Sigma providers will have an impressive list of clients, some of which are seen on other provider web sites. But some providers list clients with which they do limited work, like training a few people here and there. The provider you're looking for is one that has worked with a company at the corporate level and is the primary provider for corporate-wide Six Sigma deployments. Is the provider experienced in supporting companies your size?

Experience in Your Industry. Although Six Sigma is robust enough to be applied to all industries, finding a provider that is experienced in your industry makes sense. Qualification by similar industries may be acceptable. For instance, industrial chemical processing in batches or continuous manufacture is similar to pharmaceuticals manufacturing at the processing level. Likewise, metals fabrication often follows similar processing to industrial chemicals processing.

Thought Leaders, Customization, and Flexibility. Every Six Sigma deployment must be tailored to your company and mesh with your culture. There is no "one size fits all" deployment. The ability of your provider to quickly customize their Six Sigma training portfolio is very important. Your provider should have senior consultants who have the experience in working with several companies.

They also should be "thought leaders" in that they know the guts of Six Sigma and have the ability to deliver to you the program that makes sense to your company. Or do the providers have limited knowledge and specializations in Six Sigma? Reviewing resumes of the consultants who will likely be supporting you is an important step in the selection process. The provider should describe the process of customization in detail.

Reference Checks. References are of key importance when selecting a provider. References can vouch for the effectiveness, efficiency, and quality of the services provided by the provider. References can also be a resource to test for how well the providers transfer their technologies and enable their clients to completely internalize the Six Sigma methods.

Be sure to ask the provider for references that are strong and at least one reference that might be weak for the provider. Also, ask each reference for a list of additional people you can talk to so you don't limit your contacts to only those provided by each provider.

IP Portfolio. Because Six Sigma covers such a vast array of topics, from deployment strategy to process improvement roadmaps, the intellectual property (IP) of each provider becomes an issue. Are they equipped to give you what you need? Are there any gaps in their IP? You should obtain a comprehensive list of their IP and also see some physical examples. The references of each provider will also be a great source in addressing the quality of the provider's IP. Webex's are a valuable tool in reviewing IP with potential providers.

Bench Strength—Consultant Experience. In this popular time of Six Sigma, just about anyone who has completed training as a Black Belt considers themselves worthy to be a Six Sigma consultant. In fact, there are several CEOs of Six Sigma providers that have never been fully involved in a corporate deployment of Six Sigma.

One CEO of a Six Sigma provider drove Six Sigma into four manufacturing plants in Asia as his base experience. You would love to see a proportion of the consultants who actually drove Six Sigma into a company as a corporate leader. The alternative to this experience would be consultants that have managed corporate deployments as a senior consultant.

The other problem lies in the tendency of every Six Sigma provider asserting that their consultants are the best in the business. So, how do you know? Reviewing resumes is the first step, along with asking references if there were particular consultants that were extremely effective during the deployment. In my own company, I have always been surprised at how often the names of some of my most outstanding consultants pop up as specific requests from new clients.

You'd like to see the consulting group have a broad experience base collectively. You'll be looking for operations expertise, new product development experience, transactional process improvement experience, and sales and marketing core competencies. Also, if the provider keeps customer evaluations for each consultant, reviewing those evaluations will tell you most of everything you need to know.

One issue that presents itself often is whether the provider's consultants are employees or independent contractors. There is a sense that if the provider's consultants are independent contractors, the consulting engagement may be hampered. This should not be a big concern.

The expectations are similar for both employees and independent contractors. Sometimes the expectations are most clear to independent contractors, and their personal revenue is directly affected by how well they meet expectations. And, even consulting companies that emphasize the employee model will have a percentage of independent contractors, so you will probably never know who's an employee and who's a contractor.

The bottom line is providers that use independent contractors have lower fixed costs and are likely to be more financially healthy that those providers that use an employee model. The provider that uses independent contractors will be more flexible and have greater expertise than employee-driven companies because contractors can be hired and released with greater ease. The important issue is that the contractors are not to bill you separately from the provider. The provider should be the single point of contact with you.

Global Presence. If your company has a strong global presence, selecting a provider that can work effectively around the world becomes more important. You'll want to know if the providers have offices around the world and where. You'll also want to know into how many languages their IP portfolio has been translated. The process of accurately translating English-based materials into other languages can be time consuming, expensive, and fraught with difficulties. It is best to have the bulk of the portfolio translated at the starting gun.

QUESTIONS TO ASK PROVIDERS

The following are additional questions to include in your RFP, to have the potential providers answer in addition to the RFP, or to be included while a select few providers are doing face-to-face presentations.

Questions to ask:

1. How large is your organization?
2. How many years have you focused on Six Sigma?
3. For approximately how many clients have you completed Six Sigma implementations?
4. How many clients do you have active at this time?
5. Have you ever worked with both U.S. and European companies?
6. For a $2 billion company, what would you expect as the ongoing yearly cost savings?
7. For a $2 billion company with locations throughout North America and Europe, how many Master Black Belts, Black Belts, and Green Belts would you expect to have to keep Six Sigma going?
8. How long do you feel it would take to recover the initial cost of implementing Six Sigma?
9. How long do you feel it would take to have the program become self-sustaining?
10. Do you recommend closed (client-only) or open (public) classes when training Master Black Belts, Black Belts, and Green Belts?

11. For a three-division North American (40 locations) and European (20 locations) company, what would be a rough (range) cost for your services? This includes executive training all the way down to assisting in Green Belt training, as well as auditing.
12. Is there any area of Six Sigma that you specialize in?
13. I would like the name and number of references that I can contact. What is one that you are most proud of and one that you are least proud of?

TERMS OF ENGAGEMENT

Gain Sharing. The final contract agreed to by you and the provider will represent the nature of your relationship. A common trend today is to put a great percentage of consultant revenue on a gain-sharing agreement. The better the financial targets are met, the more money the provider makes. Although this is a relationship to consider, it can lead to trouble. First, the provider will be very concerned about the method of financial tracking you use.

The provider will probably never be happy with the way you track the money. Gain sharing can easily end up in an adversarial engagement. Also, there are many components of a deployment of which a provider has no control. That gets very frustrating for the provider. Best be clear about your expectations in the contract and leave it at that.

Selecting the Account Manager. The account manager should be treated like a senior leader in your company. In the spirit of partnership, the account manager is critical to the success of the engagement. This person should be selected before the contract is signed and delineated clearly in the contract.

Elements of the Contract. Every company has a legal department through which the contracts follow. The minimum elements should include the following:

- General description of objectives and scope of services.
- Terms of agreement.
- Who the consultant is.
- What services are to be provided.
- Where the services will be provided.
- When the services will be provided.
- Organization of personnel supplied by consultant.
- Status reports: status meetings.
- Facilities and services to be provided by customer.

- Consultant price list.
- Competitive restrictions.
- Handling of intellectual property of both parties.
- Warranties and liabilities.
- Insurance coverage requirements.
- Confidentiality between both parties.
- Termination by either party.
- Modification, amendment, supplement, or waiver.

SUMMARY

The use of external consultants will improve the efficiency and effectiveness of your Six Sigma deployment. Leveraging expertise that has been developed from participation in multiple deployments will serve to accelerate the program. Selecting the right consulting company that has the right chemistry for your company is crucial. I have seen companies lose time and results by selecting the wrong consultant at the beginning.

Selecting consults with impeccable ethics and the flexibility to match the program to your situation are two criteria you will review. Like anything else in business, good planning early on and clarity of expectations through the RFP will pay off over the short term.

5 Strategy: The Alignment of External Realities, Setting Measurable Goals, and Internal Actions

With Mike Brennan

Why even bother discussing strategy in a book devoted to the deployment of Six Sigma? Because moving from strategy to breakthrough performance is one of the toughest leadership challenges. It takes careful understanding of the external realities of your markets and your customers, setting measurable goals, and aligning your internal actions to accomplish the strategic forecasted outcomes.

In fact, in their excellent article, "Turning Great Strategy into Great Performance" (*Harvard Business Review*, July–August, 2005), Michael Mankins and Richard Steele report that, of 197 large companies surveyed, companies generally delivered 63 percent of the financial performance promised by the strategy. Many CEOs have been fired because they could not execute their strategic plan. Doing that is the real art of successful businesses.

As discussed in Chapter 2, "The True Nature of Six Sigma: The Business Model," Larry Bossidy (with Ram Charan) in his book, *Confronting Reality*, presents a unique business model that links three components:

- External realities
- Financial targets
- Internal activities

These three components are linked into a dynamic system that produces new business models with which to move your business smoothly into new markets and new products. The authors state, "Linking and iterating the financial targets, external realities, and internal activities, and searching for the right mix in each of the three components of the business model, is what determines the accuracy of the final product." You will make money by effectively linking your internal activities to the external realities and setting appropriate financial targets.

The construct that drives the linkage is the strategic plan, which is included in internal activities. Companies that realize breakthrough results with Lean and Six Sigma have strong leadership that drives the change initiative from the Corporate Strategic Plan through to the lowest levels of the organization. The total financial returns are directly related to the organizational level of the leadership driving it, as shown in Figure 5.1. Figure 5.1 displays that senior leadership driving Six Sigma as a strategic initiative will achieve noteworthy financial benefits from the deployment.

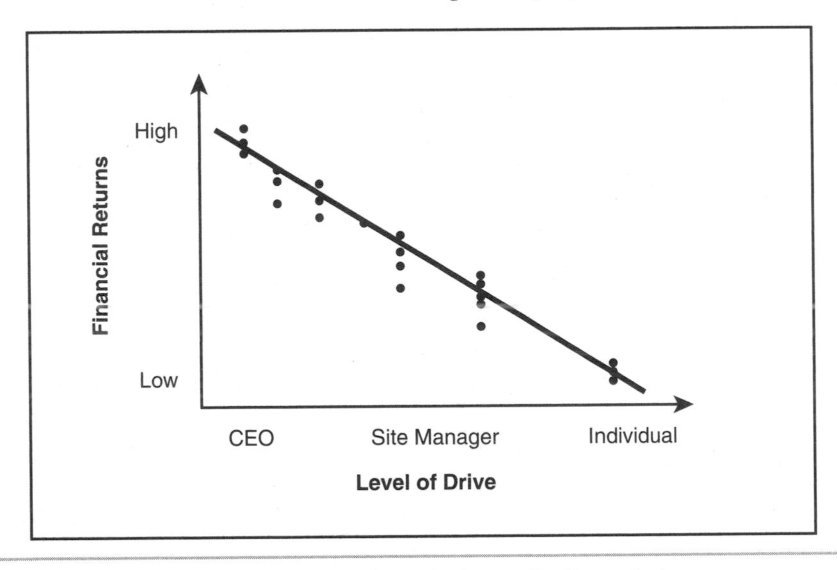

Figure 5.1 This chart shows that the financial results from a Six Sigma deployment are strongly related to the level of leadership committed to the deployment.

Six Sigma is truly a company culture change, in that it moves decision making from the political/subjective realm to a data-driven process. In their outstanding book titled *The Power of Alignment*, George Labovitz and Victor Rosansky created a linkage model that directly compares to the three-component business model.

This model is seen in Figure 5.2 and shows the vertical linkage of the corporate strategy to its people. It also displays the horizontal linkage from the business processes to the customer. In addition, the model shows that the strategy is linked to the processes, the people, and the customers (external realities). Likewise, it displays people being linked to the processes, strategy, and customers as well. This is actually a definition of Six Sigma: Optimized business processes developed by people make the customers happy. The strategy drives which customer, process, and products need to be optimized.

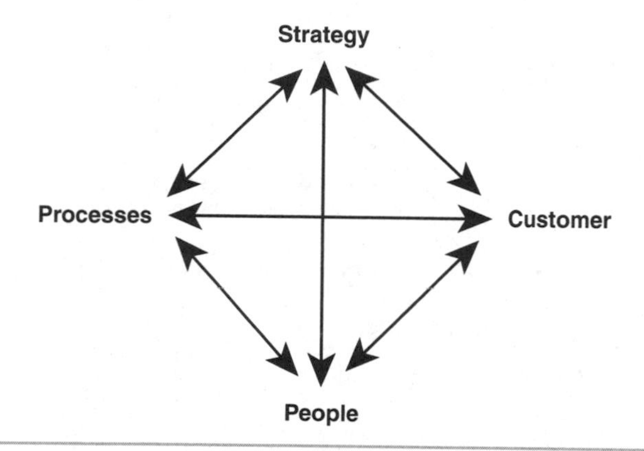

Figure 5.2 This chart shows the model developed by Labovitz and Rosansky in their book, *The Power of Alignment*. Making the critical linkages among strategy, people, customers, and processes is the key to Six Sigma success.

In fact, Total Quality Management (TQM) largely failed because organizations didn't link TQM activities to the external realities as defined in a good strategy. Likewise, reengineering largely failed because organizations didn't link processes to strategy and people.

Recently I talked to a new CEO whom I knew from my work with AlliedSignal. I asked him how he was doing in his new job. He said it was easy: "Just improve growth and productivity." He was right. Of course, growth is very difficult if your company's productivity is inadequate. To grow, your operational excellence must be there. Regardless of whether your company has a strategic plan or not, the first step in the improvement process is to move to operational excellence in your existing business lines. You shouldn't waste time in getting started, as every day of delay is lost profit. Most companies, in fact, do start Six Sigma by focusing on productivity in one form or another.

The data analysis of more than 50 different companies and their Six Sigma deployments shows that 1 percent of your population of employees trained as Black Belts can deliver four profit improvement projects per year, each with an average profit improvement of $250,000 per project, for a total of $1,000,000 per trained full-time Black Belt per year. In the first two years of deployment, you should expect to achieve 2 to 4 percent of your revenue in profit improvements. From a timeline standpoint, getting your existing business to operational excellence is the first order of business, and Mehrdad Baghai in his book, *The Alchemy of Growth*, calls this Horizon 1, as shown in Figure 5.3.

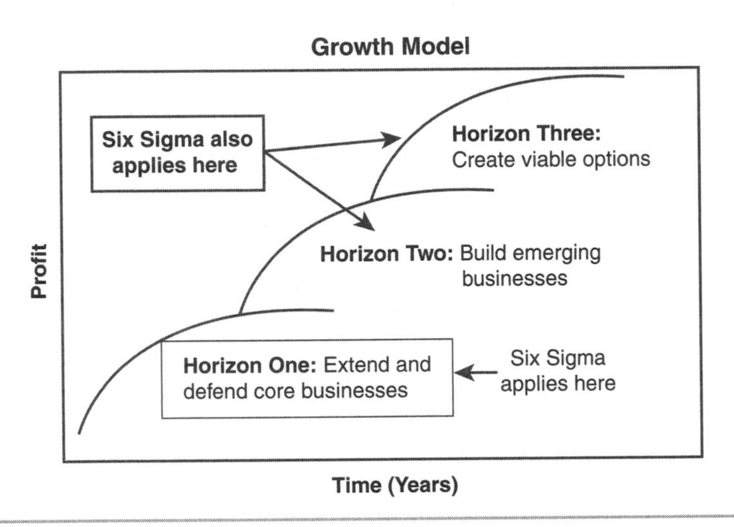

Figure 5.3 Adapted from Baghai, Coley, & White. The three horizons of growth. Strategic plans should include plans to move further in each horizon.

Baghai's model consists of three growth horizons. The first horizon (extend and defend the core businesses) is focused on shoring up the core businesses and getting the potential growth that remains. By successfully executing Horizon 1, you will earn the opportunity to start Horizon 2 and 3 growth activities. So Horizon 1 tends to be focused on productivity as a priority, with growth as a secondary priority.

The second horizon (build emerging businesses) places a premium on growth and building new streams of revenue, rather than improving productivity. Horizon 2 focuses on the new potential star products or services of the company and extending the core business. Horizon 3 (create viable options) is focused on creating new business out of research or pilot projects and creating new core businesses. Horizon 3 is the lifeblood of a long-term company.

Taking the concept of improving productivity and then focusing on growth, improving Horizon 1 is the ticket to creating strategic actions in Horizon 2 and Horizon 3. You have to be able to eat short term to grow long term.

An excellent strategic plan will include ideas and actions for all three horizons. Six Sigma is ideal for Horizon 1 because the priority is improving operational excellence.

But Six Sigma is also instrumental in Horizon 2 and Horizon 3, with Design for Six Sigma, Technology for Six Sigma, and Marketing for Six Sigma. Therefore, the strategy can yield Six Sigma projects for all three horizons simultaneously.

The Six Sigma project selection process starts with the measurable corporate business goals. The key word here is "measurable." If the goal isn't measurable, with a baseline value of where you are today and a value of where you want to be at a specific time in the future, then you don't have a goal. Each business unit identifies its own measurable goals that support the corporate business goals. These measurable goals are called Business Critical Ys to represent the output from the completed projects. Then, project cluster areas are identified, like accounts payable, working capital, supply chain, and so on. From here, individual projects can be identified that are linked to the business goals, as shown in Figure 5.4.

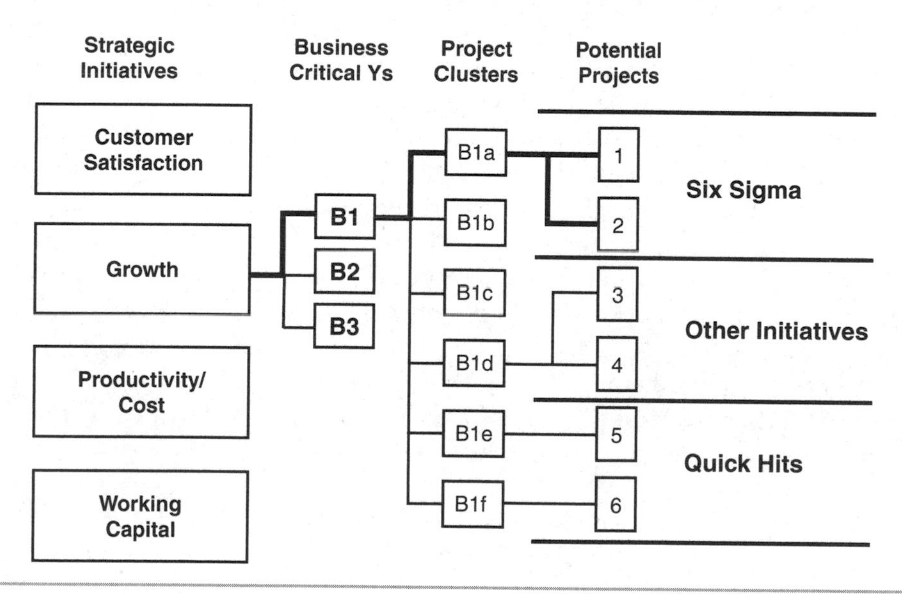

Figure 5.4 The linkages linking the top strategic initiatives to the measurable Critical Ys, to project clusters, and finally to projects. Projects all have a "line of sight" to at least one or more strategic initiative.

Now that we have kicked off projects that will improve our base business, it is time to turn our attention to the medium-term strategic goals, or from a timing standpoint, Horizon 2, as shown in Figure 5.5.

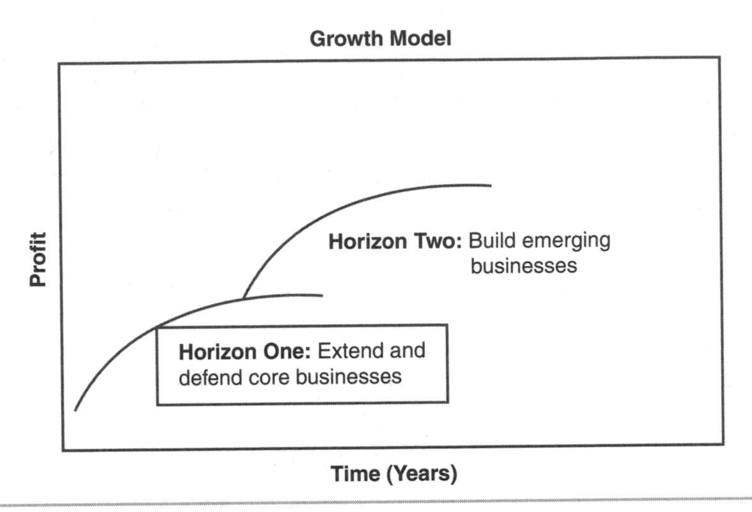

Figure 5.5 Horizon I success is necessary to earn growth in Horizon 2.

Even though Motorola started Six Sigma, it was General Electric that showed the world on a large and dramatic scale that the Six Sigma methodology applies to any business process. As we have seen earlier, the basic equation of Six Sigma is: the Output (Y) is a function of some number of input variables (Y is a function of Xs). This approach can also be applied to the strategic planning process. Figure 5.6 shows that the typical financial outputs of a company are a function of certain input variables.

Figure 5.6 displays some of the financial inputs for the outputs (Ys) of return on investment, revenue, profits, cash flow, and economic value add. For example, we can say that revenue is a function of quantity sold, average selling price, and discounts. Of course, there are a lot of other inputs for revenue, but the point is that some inputs are more important than others. The process of Six Sigma is designed to identify the most important inputs, optimize those inputs, and then control them.

Functions We Want to Maximize
Lend Themselves to Six Sigma Analysis

Y Return On Investment = *f* (Profits, Investments)

Y Revenue = *f* (Quantity Sold, Average Selling Price, Discounts)

Y Profits = *f* (Revenue, Product Cost, Business Costs)

Y Cash Flow = *f* (Profits, Working Capital)

Y Economic Value Add = *f* (Profits, Capital Charges)

Figure 5.6 The Critical Ys are a function of inputs. Projects improving those inputs will improve the performance on the outputs.

Figure 5.7 shows the roadmap we will use in the strategic planning process. The first step is to discover the financial factors that determine the long-term value of a company. Then we look to the external realities to determine the specific industry-critical drivers of competition. Next, we identify the gaps in our company's competitive position and opportunities to exploit. To address the gaps and opportunities, we pick a balanced portfolio of cumulative fractional improvements to better position ourselves. The last step is the one where most companies fail to achieve their strategic intent, and that is the execution of a comprehensive deployment plan.

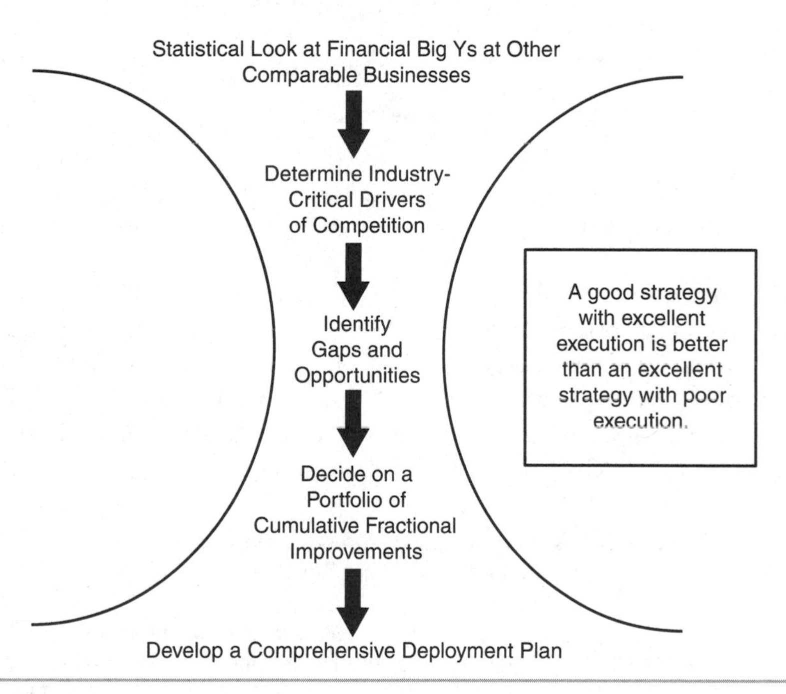

Figure 5.7 This is one roadmap leading to the output of a comprehensive strategic deployment plan. The strategic plan addresses the external realities by aligning internal actions to the strategy.

General Electric did a benchmarking study identifying factors that have the biggest influence on long-term company value. As one might expect, return on investment (ROI) has the biggest effect on long-term company value, as shown in Figure 5.8.

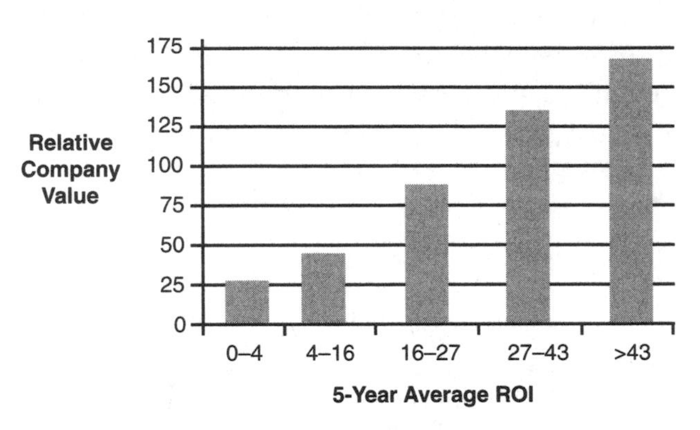

Figure 5.8 The linkage of strategic drivers to the long-term company value. The more strategic drivers that are made real to the company, the more value the company delivers.

GE also did an analysis on what factors affected ROI. Figure 5.9 shows "the more, the better factors" like market share (be #1 or # 2 in your market), relative product/service quality (you have preferred product or service), and so on. It also shows the factors that are trade-offs between short-term profitability and long-term value. The key finding was that businesses that perform best over the long term are more aggressive in investing in marketing, R&D, and capacity expansion than competitors.

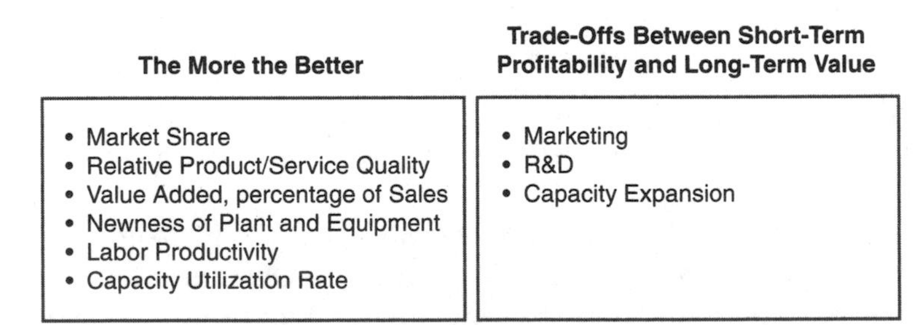

Figure 5.9 Businesses that perform best over the long term are more aggressive in the trade-off factors of marketing, R&D, and capacity expansion than competitors.

Now our job is to find the drivers of competition, as shown on Figure 5.10, which will allow us to be the most preferred supplier by our targeted customers. Being the most preferred supplier of products or services drives revenue growth. Once we have generated the revenues, the operational drivers will determine how effectively and efficiently we deliver on the drivers of competition. And finally, the financial drivers will determine how well we achieve our business goals.

Finding the Right Strategic Drivers is the First Step

Strategic Drivers

Drivers of Competition
?
?
?
?

Operating Drivers
Product Planning
Pricing
Market Share
Operating Expense
SG&A Expense
Goodwill
Property & Equipment
Working Capital
Debt
Equity

Financial Drivers
Sales
COGS
SG&A
Taxes
Invested Capital
Cost of Capital
After-Tax Profit
Capital Charge

Return on Investment

Figure 5.10 Identifying the factors that drive competition are linked to strategic and operational drivers and then to financial targets.

Interviews with your customers will reveal their needs and how your company compares to your competitors in delivering on these prioritized customer requirements. Design for Six Sigma's (DFSS's) Quality Function Deployment (QFD) is used as a planning tool. The right side of the QFD House of Quality depicts a graphical representation of how you fare against competition in meeting the prioritized customer requirements. Figure 5.11 shows an example company. A set of customers rated the company on several drivers of customer satisfaction to include in order of importance and the customer rating. For the customer rating, the higher the rating, the better:

1. Product Quality (3)
2. Brand (1)
3. Product Performance (6)
4. Delivery (6)
5. Product Offerings (3)
6. Service (6)
7. Relationship Management (6)
8. Ease of Doing Business (5)
9. Marketing Support (8)
10. Price (7)
11. Value (4)

Product quality is the most important requirement the customer has, and the customer rates the example company as a 3 out of 9, or inferior to competition. Even worse, the customer perceives the brand name as totally inferior. In addition, this company's product offerings are scored low. This analysis, using direct customer feedback, gives the company three clear areas in which to work.

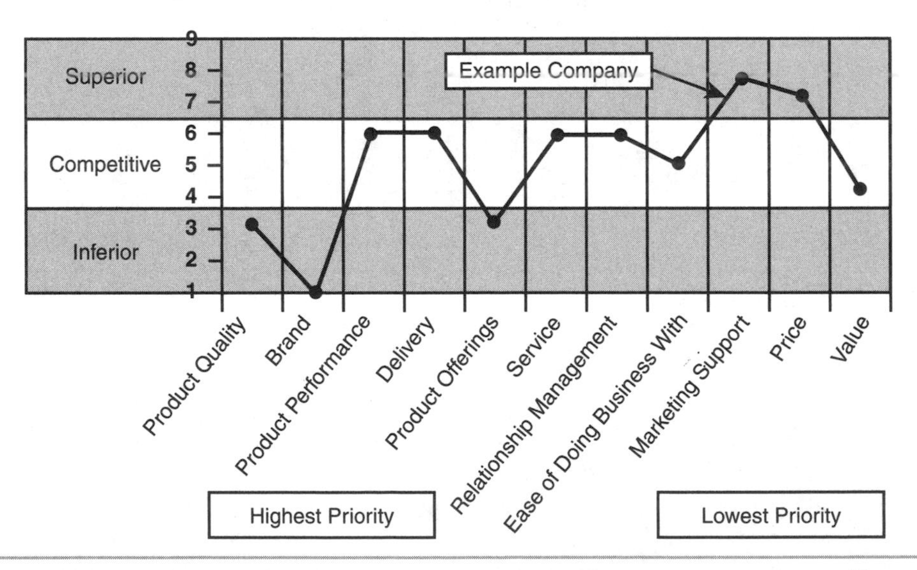

Figure 5.11 Customer rating of a company's competitive drivers. This company has been rated low in product quality, brand, and product offerings. These are good drivers to address with a strategy.

In contrast, Figure 5.12 shows the competition as superior in the top three of five important customer requirements. However, the competition is inferior in several areas including service, relationship management, and ease of doing business with. So, this company has a very clear picture of its external realities, and the next steps are to align a strategy with these realities and to align its internal actions with the strategy.

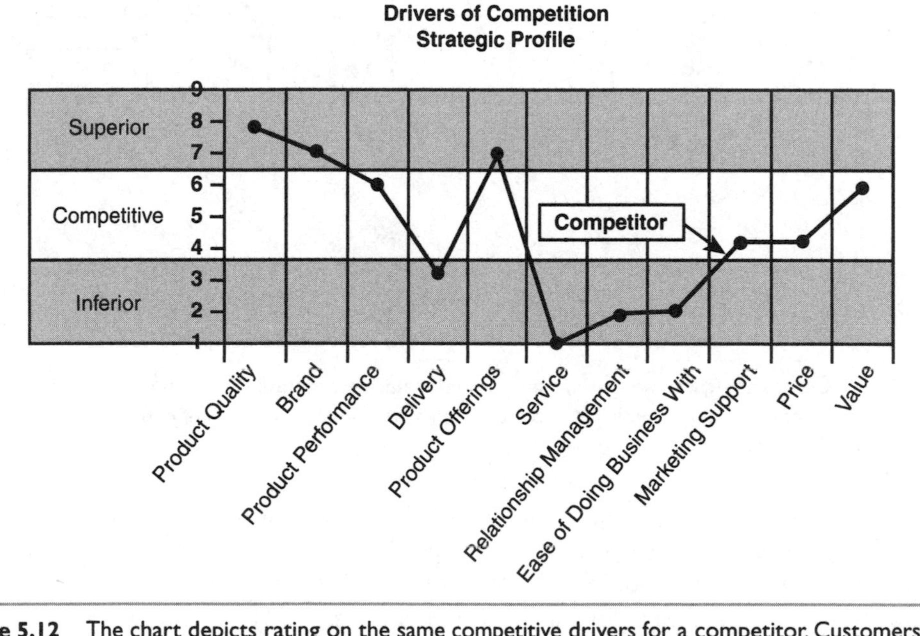

Figure 5.12 The chart depicts rating on the same competitive drivers for a competitor. Customers rated this competitor low in delivery, service, product offerings, relationship management, ease of doing business, and marketing support.

It is time for this company to do a gap analysis. Contrasting the example company to its competition in Figure 5.13, we get a clear picture of the gaps and opportunities the example company has to address. Our company is weak in product quality and strong in customer service. This is useful strategic information.

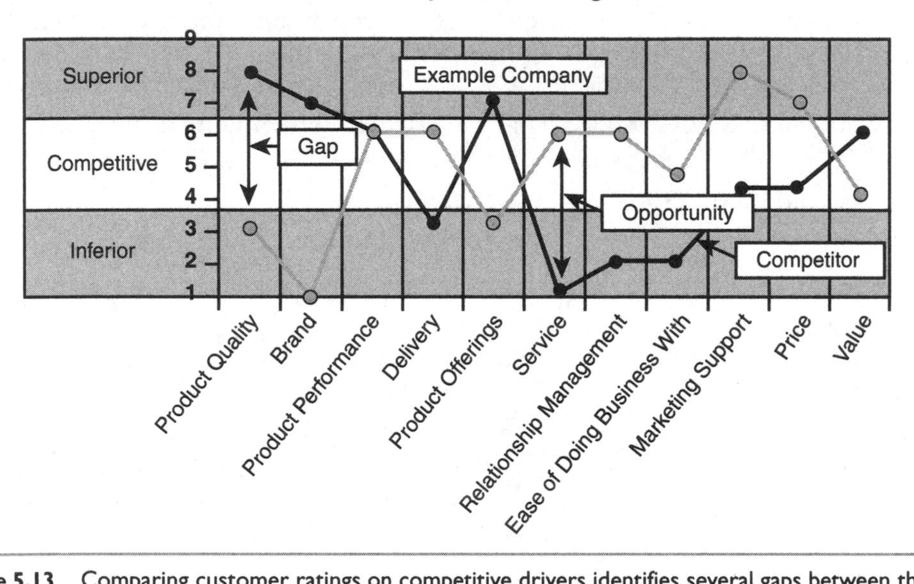

Figure 5.13 Comparing customer ratings on competitive drivers identifies several gaps between the company and competitors. Some of these gaps will have actions defined that will close those gaps.

Following our strategic planning roadmap of Figure 5.7, now that we have looked at gaps and opportunities against our competitors, we can identify other opportunities.

Figure 5.14 shows a potential set of options to consider in your strategic planning. These options are classified according to two categories: (1) The business return on each initiative and (2) Degree of business, product, or technology change. The temptation here is to try to do too many breakthrough initiatives at once. It is better to have a balance of short-, medium-, and long-term initiatives, shown in Figure 5.15, which we will refer to as a balanced basket of fractional improvements.

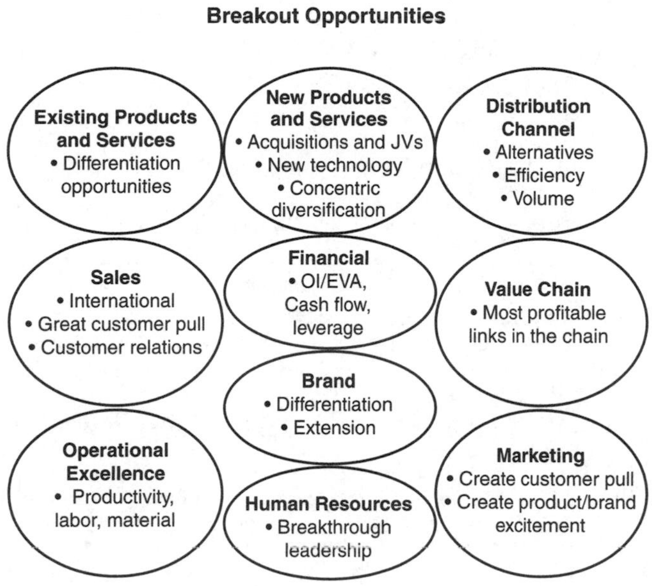

Breakout Opportunities
Existing Products and Services
• Differentiation opportunities
New Products and Services
• Acquisitions and JVs
• New technology
• Concentric diversification
Distribution Channel
• Alternatives
• Efficiency
• Volume
Sales
• International
• Great customer pull
• Customer relations
Financial
• OI/EVA, Cash flow, leverage
Value Chain
• Most profitable links in the chain
Brand
• Differentiation
• Extension
Operational Excellence
• Productivity, labor, material
Human Resources
• Breakthrough leadership
Marketing
• Create customer pull
• Create product/brand excitement

Figure 5.14 Clusters of breakout opportunities based on gaps in strategic performance.

Maximize Returns and Minimize Risks by Choosing a Balanced Basket of Fractional Improvements
Breakthrough
V
IV
III
Substantial
II
I
Incremental
High
Low
Return on Individual Initiatives
High
Low
Degree of Business/Product/Technology Change
Initiatives and Effort Required
*Adopted from Steven Wheelwright model

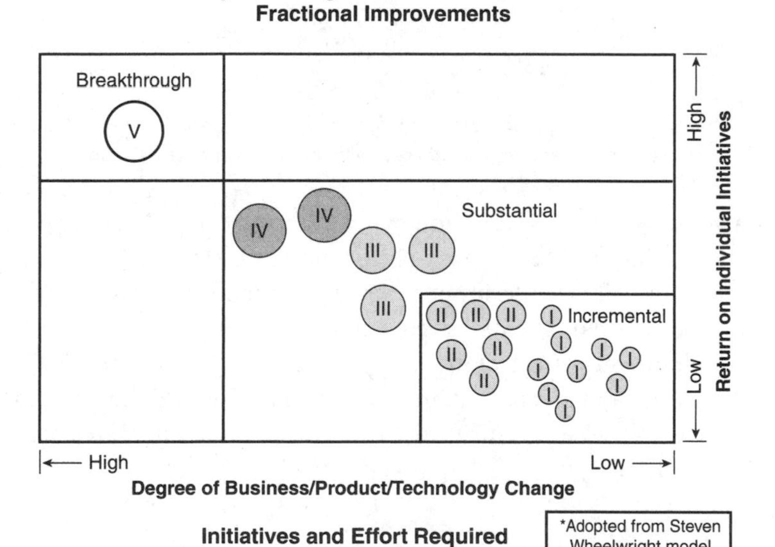

Figure 5.15 Categorizing strategic opportunities by degree of business impact and potential return.

Now that we have funneled down to a set of concise objectives, the strategy deployment plan is how you will achieve the results. The plan and tracking phase is displayed in the roadmap shown in Figure 5.16.

Strategic Planning Activities

External | Internal

Situation Analysis
• Industry/Environment
• Financial Benchmarks
• Competition
• Suppliers
• Channels
• End Users
• Organization
• Financial
• Marketing
• Sales
• R&D
• Operations
Data Gathering
What Do We Know?

SWOT
• Each Function
• Company Competitors
• Critical Players
What Do We Need to Find Out?

Objective
• Specific • Measurable
• Challenging vs. Benchmarks
Identify Opportunities

Strategies
• Overall Approach
• Use Leverage from SWOT
• Alternatives

Plans
• Who • What • When
• Resources • Communications
Success is in Execution

Tracking
• Targets Responsibility Key • Measures Quantified
• Timing of Reviews

Figure 5.16 Roadmap for activities associated with strategic planning, starting with the external and internal realities and ending with targets, accountability, and review timing.

The key to achieving these results is following a strategic deployment plan, which includes communicating the plan, setting performance objectives, and putting in place the infrastructure that will provide support and accountability, as shown in Figure 5.17. This is the process you will use to effectively deploy Six Sigma. Six Sigma must be linked to the strategy deployment plan. The most important box in the Figure 5.17 graphic is "Developing Objectives and Measurement Criteria." The measurable objectives you define become the Critical Ys that start the top-down project selection process.

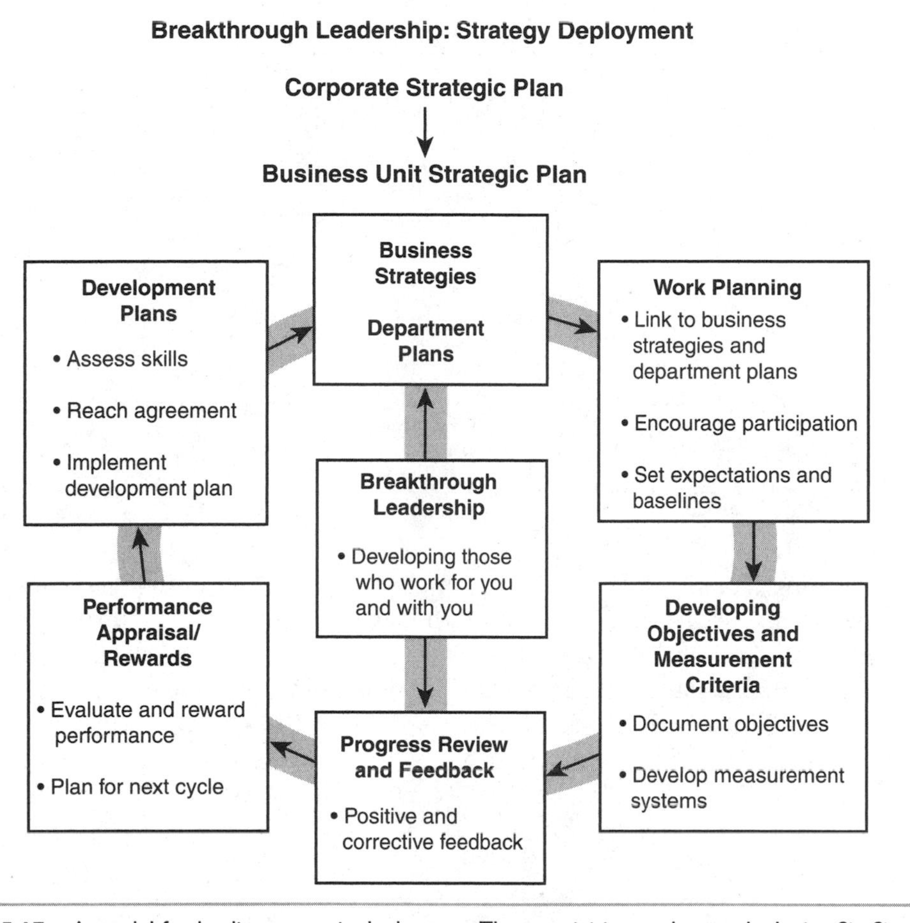

Figure 5.17 A model for leading strategic deployment. These activities are key to deploying Six Sigma.

Just as building a solid base business in Horizon 1 is the price of admission to Horizon 2's growth, the success in Horizon 2 will provide the opportunity for longer-term strategy deployment in Horizon 3, as shown in Figure 5.18.

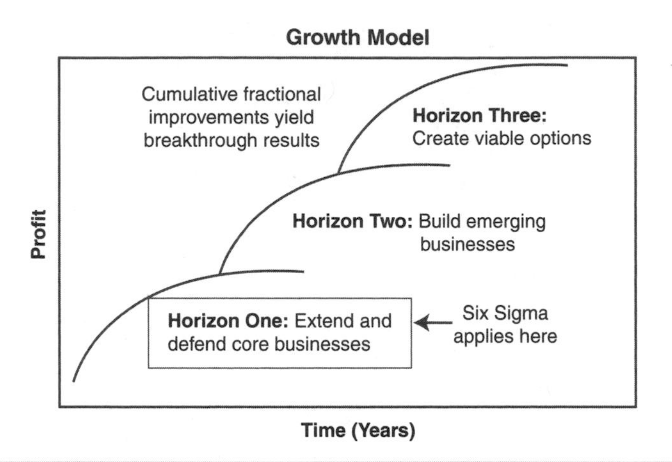

Figure 5.18 The three horizons of growth are achieved by executing a series of fractional improvements in the right places.

The following chapters will guide you through the process of planning and executing on Horizon 1 and perhaps even initiatives addressing Horizon 2 and 3. The most difficult task of any company is to define a viable strategy and execute it systematically. Deploying Six Sigma and other strategic initiatives will help your company do that. To use the strategy to guide Six Sigma project selection is critical. Figure 5.19 shows how this works.

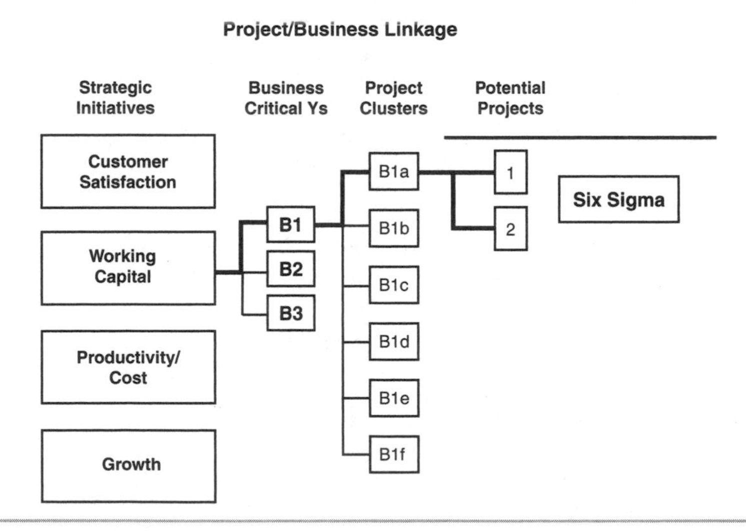

Figure 5.19 Strategic initiatives linked to Business Critical Ys, project clusters, and projects. Each of the four strategic initiatives would have a similar tree alignment.

Going from the strategic initiatives creates measurable objectives we'll call Critical Ys. The Critical Ys each create a set of project clusters. The organizations then identify projects for each project cluster. Before you know it, you could have hundreds of projects in progress, each addressing important aspects of executing your strategy. The rest of this book gives you the details on how this happens.

6 Defining the Six Sigma Program Expectations and Metrics

With Joe Ficalora

You've assessed your external realities and have a clear understanding of your company's gaps in performance with your customers. You've done the research and decided that Six Sigma is the next major initiative. Now, before your official launch of Six Sigma, your leadership team must develop a clear and easily understandable set of expectations and metrics for the new Six Sigma program. This is to certify that internal actions associated with Six Sigma will be linked to financial targets and corporate strategies.

The metrics are sometimes the tricky part because sometimes we track the metrics of interest and sometimes we don't. For example, a prominent metric associated with Six Sigma is the estimate of a process's rolled throughput yield (RTY)—the percent of work units that make it through the process error-free. The RTY of an invoicing process of 80 percent means that 80 percent of the invoices processed make it through the process without any defects. Yet, 20 percent of the invoices are found to have one or more defects. Because invoice accuracy is closely tied to customer satisfaction, that 20 percent is a big number for large companies.

Not all the metrics are financial in nature, but they are ultimately linked to financial targets. For example, in 1995, my $4 billion sector of AlliedSignal had a financial target of $75 million to $125 million in pretax income. But we had other metrics and goals as well. The base operational metrics were rolled throughput yield (RTY), cost of poor quality (COPQ), and capacity productivity (C-P). Table 6.1 shows a summary table of these three metrics and how they were tracked quarter to quarter.

Table 6.1 An Example of Three Six Sigma Metrics, Goals, and Actual Performance

Key Metrics	1994 Baseline	1995 Goal	1st Quarter Goal	1st Quarter Actual	2nd Quarter Goal	2nd Quarter Actual
RTY	77%	82%	79%	80%	81.3	79.4
COPQ	19%	15%	17%	17%	16%	14.5%
Capacity	1	1.14	1.09	1.08	1.13	1.13

This table shows the three Six Sigma metrics. The starting points for each metric were the previous year (1994) baselines. The goals for 1995 were set for the year and prorated for each quarter. Actual results were reviewed each quarter. While none of these are true financial targets except parts of cost of poor quality, they are each linked to financial targets. RTY is directly related to margins because the fewer defects you produce, the better the margins. COPQ is directly related to margins because the less you lose due to waste, the better you perform. If the capacity goal is met (moving from baseline to adding 14 percent capacity), then capacity is related to additional revenue if your factories are capacity constrained (which ours were).

There were other Six Sigma metrics as well, but these three were the top metrics for our factories. Other Six Sigma metrics tracked were safety, customer satisfaction, materials management, and inventory. We ended up beating all the sector goals and our financial resources put the final number at $95 million pretax income for the year. That's about equivalent to adding a $950 million business to the sector without any new people or capital. And these were the results that are reported directly to Wall Street.

Another large company (about $36 billion) set these three goals for Six Sigma:

1. Growth: $9 billion in additional sales.
2. Cost: Three-year saving = $1.0 billion.
3. Cash: $3.5 billion.

These goals would total more than $13.5 billion in three years. Talk about creating a positive sense of urgency.

But the important thing is to set aggressively clear goals because if you shoot low, you'll hit it. Only by aiming high can you expect excellent results that will transform your company. My company worked with a large $7 billion company. By the end of the first year, I discovered the CEO's goal for Six Sigma was $12 million. For a company that size, that goal was embarrassingly low. So, either the CEO didn't understand the potential benefit of Six Sigma or was just doing Six Sigma to say he was doing it. I couldn't help myself. I sent this CEO a letter.

Mr. John Doe
Chief Executive Officer
InfoSystems

Dear John,

We are closing on the second year of Six Sigma deployment, and InfoSystems has come a long way in moving toward becoming a Six Sigma company. We have enjoyed our relationship with InfoSystems, and Christine, her staff, and the operations folks have been a pleasure to work with.

I am certain that you will meet your 1999 financial goals of $12 million for the program. I would like to share some results from some of our other clients and suggest a more aggressive goal for InfoSystems for the year 2000.

One of our smaller clients (about $2B in sales) will bring in **$50M+** from 200 Black Belts and Green Belts averaging about $200K per "Belt." We've just launched a chemical company ($1.2B in sales) whose CEO is a former AlliedSignal business president. This CEO set a **$50M** goal for the year 2000 based on the results he obtained at AlliedSignal.

We are launching another $7B company whose COO is setting a goal of **$100M** for the year 2000. Another $4B company is setting a **$75M** goal. During my work with the $4B AlliedSignal engineered materials sector, we achieved $95M of pretax income the first year, $110M the second, and $160M the third.

John, frankly I think your 1999 goal was low. I don't think your goal created the sense of urgency that is essential to make operational breakthroughs in your company. I recommend you announce soon that your expectation for 2000 is **$100M in pretax income**. You'll see a totally different dynamic occur, and I think this goal will stimulate commitment and activity at a level you haven't experienced before.

You move into 2000 with 150 trained Black Belts, some Green Belts, and several Master Black Belts and Master Black Belt candidates. With the continuation of Six Sigma training at the same level, you should have resources that, when brought to bear around an aggressive goal, will achieve results that will have a major impact on InfoSystems. A $100M goal equates to about 400 $250K projects. That is about two projects per Black Belt if you continue to train aggressively. Your Black Belts should be averaging between $250K and $750K per year. My consultants report that many of your Black Belts are attaining that level of performance.

Successful Six Sigma implementations have always been marked by the concept, "If you think big, you'll get big." Please consider my suggestion seriously. I want InfoSystems to be one of the few classic case studies demonstrating real breakthrough in performance using Six Sigma.

I'd love to discuss these ideas with you in the near future. Just let me know and I'll be glad to visit.

Yours truly,

Stephen A. Zinkgraf, Ph.D.
Chief Executive Officer
Sigma Breakthrough Technologies, Inc.

Well, John easily achieved his $12 million goal and reported it to the Wall Street analysis. He did raise the goal for the next year to $56 million and got pretty close to that one. If he'd stayed at the $12–$15 million level, he would have lost about $40+ million for the year. In the Six Sigma business, it pays to think big.

DEFINING THE BOTTOM-LINE SIX SIGMA PROGRAM EXPECTATIONS AND METRICS

It is crucial that each leader set the priorities for the Six Sigma program *before the outset or launch*. Bottom line (productivity improvement), top line (revenue improvement), or customer satisfaction and participation targets need to be set and prioritized. Any Six Sigma program is distinguished by its results, and those results must be quantifiable. These results can be separated into expectations for the bottom line, the top line, and participants too. In the prelaunch phase, Executive and Champion workshops are completed to establish the priorities, plan out the program expectations, and define the strategic links and metrics. Don't launch your deployment until the hard work of preparing the launch plan is done. This is one place where Ready-Fire-Aim will not work. Taking the time to aim the program at the right targets will pay off during the launch of the program.

BOTTOM-LINE EXPECTATIONS—PRODUCTIVITY AND EFFICIENCY IMPROVEMENT

Without having experienced it personally, the numbers from Six Sigma at first seem unbelievable. That efficiency can be improved enough to generate the cash discussed next seems unrealistic at first. However, if you stop and think about how many chronic problems you have within your company that seemed too complex at first glance, you will arrive at a fairly good project list. Review this list, and you will get an idea of how much awaits you down the road to Six Sigma.

Financial Metrics. You will expect a return of about 1 to 3 percent of gross revenues in profit during the first 24 months of deployment. Setting such an aggressive goal requires determination and clear commitment from everyone in the organization. The leader can partition this goal between business units and functional groups using goal trees.

Here's a $1B dollar company example: Goal is anywhere from $100M to $300M bottom line. At $250K/project, this is between 400 and 1,200 projects, depending upon size and scope. Each BB must do 3–4 projects/year. The training target to create the critical mass of Black Belts would be between 150 and 300 BBs trained over the first two years. Training Green Belts will add increased $$$. Figure 6.1 displays what this process would look like as a roadmap.

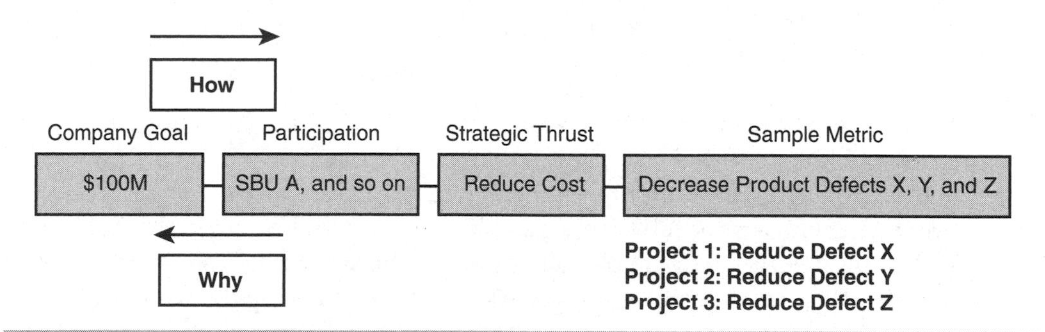

Figure 6.1 A $1 billion company sets a goal of $100 million. This is an example goal tree that shows the generation of three projects for one strategic thrust.

All bottom-line expectations are tied to dollars in one way or another. Dollars are often tied to labor and materials, but only at the project or lowest levels. At the deployment level, it is best handled with a goal tree linked to the strategic focus areas. Goal tree strategic categories and metrics are as follows:

- Productivity
- Operating income
- Delivery/speed
- Working capital, days sales outstanding (DSO)
- Cycle time reduction
- Health, safety, and environment
- OSHA reportable incidents
- Safety audit findings
- Environmental fines and warnings
- # Hazmat spills and calls
- Customer satisfaction/warranty
- Returns and warranty claims
- Satisfaction ratings
- Surveys
- Repeat purchases
- Percent new customers
- Growth
- Sales revenue
- Percent market share
- New market penetration

General Managers sometimes ask: *"How many people will be reduced from the payroll?"* By expecting this link, there is a traceable path to labor cost reduction claims. This "macro" view is unfortunately not fine enough to judge each Six Sigma project. The savings must be counted project by project and linked to key metrics that ultimately determine the business operational costs. The metrics come from the goal trees. Labor savings is only one part of the equation, and will eventually show up as business efficiency. Reducing head count might be a result of increased efficiency, but should not be a driver of it.

Key Metric—Defects per Unit. The traditional early Six Sigma metrics are related to defects and the causes of scrap, and rework. Eliminating defects directly improves productivity and customer satisfaction. By eliminating defects, cost and schedule variances improve. Motorola proved this with their original Six Sigma program. The general responses to the question, "How do I know I have defects?" usually lead to investigating nonconformance to expectations or requirements. Defects typically cause the following:

- Customer dissatisfaction and complaints with product or delivery
- Warranty claims or returns
- Rework or scrap or variance requests
- Late/early shipments

Not all defects are created equal. Defects are prioritized through the use of Pareto's addressing frequency, count or occurrence rate, cost to business, and impact to customer. Defects are the key root causes to poor productivity and mediocre customer satisfaction. Defect rates of occurrence directly relate to whether or not an operation tracks defects rigorously on a day-to-day basis. If defects are not rigorously tracked, there is no way their type and frequency are known. And there's no systematic effort to eliminate them.

Most operations, if they track defects at all, will track them at process testing stations or at final product test. The weakness of that approach is that most of the defects that occur are not considered. If an operator creates a defect and then fixes it, that defect will never show up in the system.

The common metric for tracking defects is defects per unit. The formula is DPU = (Total Defects/Total Units Processed). If DPU is chosen as a metric, most organizations will require the establishment of a defect tracking system. Within the first year of Six Sigma at Motorola, almost all operations had installed a defect tracking system. Just by doing that and making defects visible, a noticeable improvement in operations was immediately apparent.

DPU is also a component of a more robust metric, rolled throughput yield (RTY). RTY represents the percentage of units processed that had no defects. The calculation is pretty straightforward. Figure 6.2 shows the First Pass Yields of each step in an actual automotive tire manufacturing line. The First Pass Yields are equivalent to DPU, but is usually oriented more toward testing results.

Process Step	First Pass Yield
Mixing	98%
Calendar	93%
Tread Extrusion	79%
Curing Presses	97%
Spray Booths	99%
Sidewall Buffing	96%
Uniformity	88%
Sort and Label	94%
Steel Belt Cutters	98%
Body Ply Cutter	99%
Tuber	93%
Sidewall Extrusion	91%
First Stage Building	99%
Second Stage Building	99%
RTY	44%

Figure 6.2 The First Pass Yields for each process step when manufacturing car tires and the resulting RTY.

This chart shows that, except for one step, the yields are well over 90 percent. But, when you inspect RTY at the bottom of the chart, you see that only 44 percent of the tires go through this process without a recorded error. That means a huge amount of time is spent on fixing tires instead of shipping tires. RTY is calculated just by multiplying the First Pass Yield at each process together (.98 × .93 × .79....). And you can clearly see that the first place to work would be the Tread Extrusion process, the process with the worst yield.

During the early years at Motorola (the late 1980s), every manufacturing line could review and report the data listed in Figure 6.2 almost in real time. Very few companies have that capability. The ability to document defects in real time and investigate their structure to drive improvement activities was a way of life. Common defects in several functions are listed after Figure 6.2. Needless to say, DPU is a terrific operational metric.

But most companies will be required to implement a defect tracking system before DPU can be effective. These systems are simple, straightforward, and not difficult to implement. The hard part is getting everyone to put in every defect they encounter.

Figure 6.3 shows a typical trend chart for DPU. In this case this chart documents defects per 1,000 units. The Six Sigma goal line is also shown, and you can see that defect elimination projects do not seem to be working well, but there is a nice downward trend toward the end of the time period. This chart is similar to those used in Motorola.

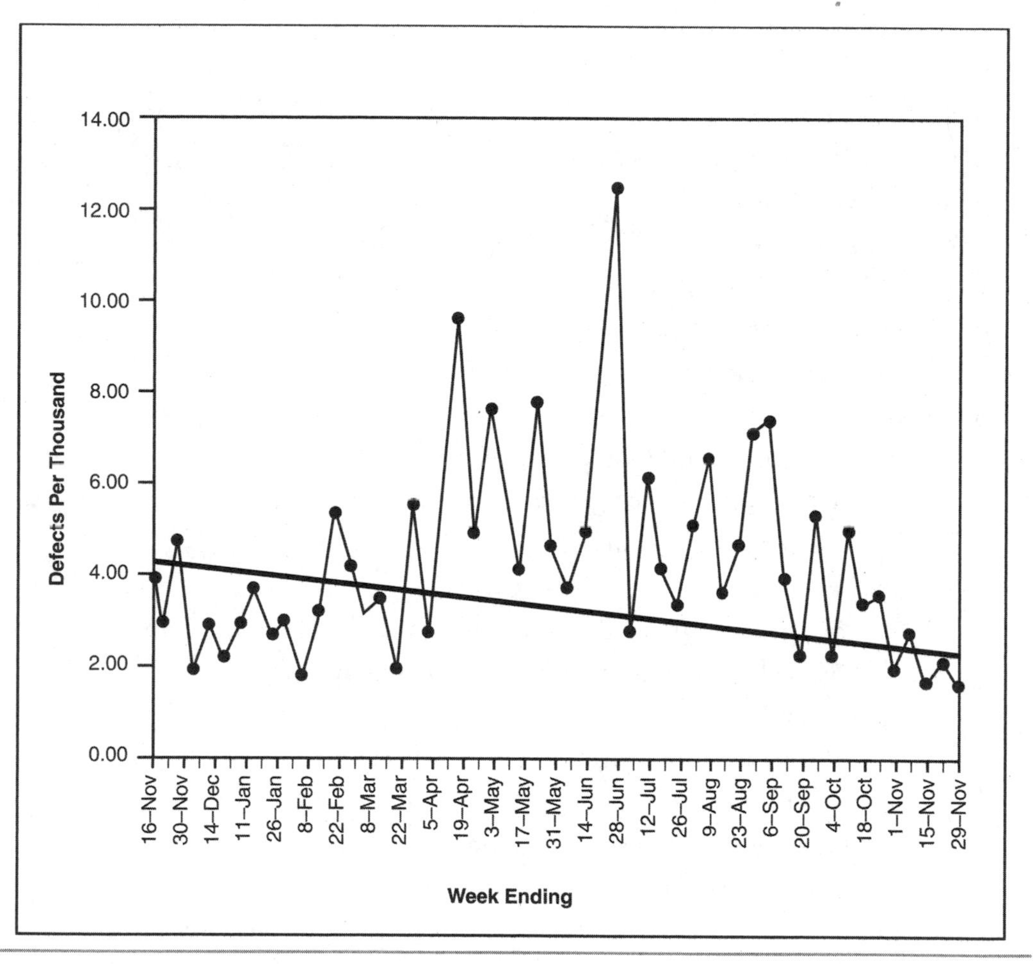

Figure 6.3 Example trend chart for defect per unit with a goal line.

1. Some Manufacturing Defect Examples:
    a. Burrs or chips
    b. Scratches
    c. Component placement
    d. Contaminants, dirt, stains
    e. Mislabels, missing labels
    f. Wrong color, wrong options
    g. Wrong packaging
    h. Missing items
    i. Supplier base defects
2. In General Business Process Defects:
    a. Missing information
    b. Missing signatures
    c. Wrong information and typos
    d. Wrong routing of information
3. In Service Business Defects:
    a. Banking and finance
        i. Missing info on mortgage application
        ii. Wrong info on mortgage application
        iii. Missed signatures in sign-off loop
    b. Healthcare
        i. Incorrect diagnosis due to missing patient information
        ii. Missed allergy notations resulting in wrong antibiotic
    c. Transportation
        i. Wrong airline luggage tag
        ii. Overbooking seats
        iii. Missing special meal annotation
        iv. Missed arrival times
    d. Utilities
        i. Wrong meter reading
        ii. Incorrect billing statement
        iii. Misapplied credits

e. Auto repair
    i. Unnecessary repairs due to wrong diagnosis
    ii. Missed warranty coverage (customer pays instead of manufacturer)
f. Restaurants
    i. Wrong food order
    ii. Incorrect food bill
    iii. Unavailable selections

4. In New Product Introductions Defects:
    a. Errors on drawings or prints
    b. Missing information on drawings
    c. Late release
    d. Missing start-up information
        i. New capital equipment omissions
        ii. New manufacturing or supplier capability omissions
        iii. New supplier qualification omissions

Key Metric—Scrap. This metric is sometimes a good place to start because most MIS systems will at least capture scrap in a manufacturing environment. This is harder to measure in a business process and may not be readily available in those situations. But scrap is surprisingly of sufficient magnitude to adversely affect productivity. Here are some examples of things to measure in the category of scrap:

1. In Manufacturing:
    a. Final test rejects in assembled products
    b. Rejected batches in chemical industries
2. In Business and Service Processes:
    a. Rejected applications
    b. Rejected forms and submittals
    c. Missed rate lock for mortgage
    d. Special request denials
    e. Wrong repair done on a car
    f. Incorrect installation of copy machine
    g. Food refused at a restaurant

3. In New Product Introductions:
   a. Product recalls
   b. Market flops
   c. Product cancellation

Figure 6.4 shows a typical trend chart for scrap over a given period of time. There is also a goal line shown. You can observe the effectiveness of reducing scrap quickly. For this operation, scrap is high and stable. The efforts to reduce scrap have not been implemented well or designed well. But, at least they know and can make adjustments.

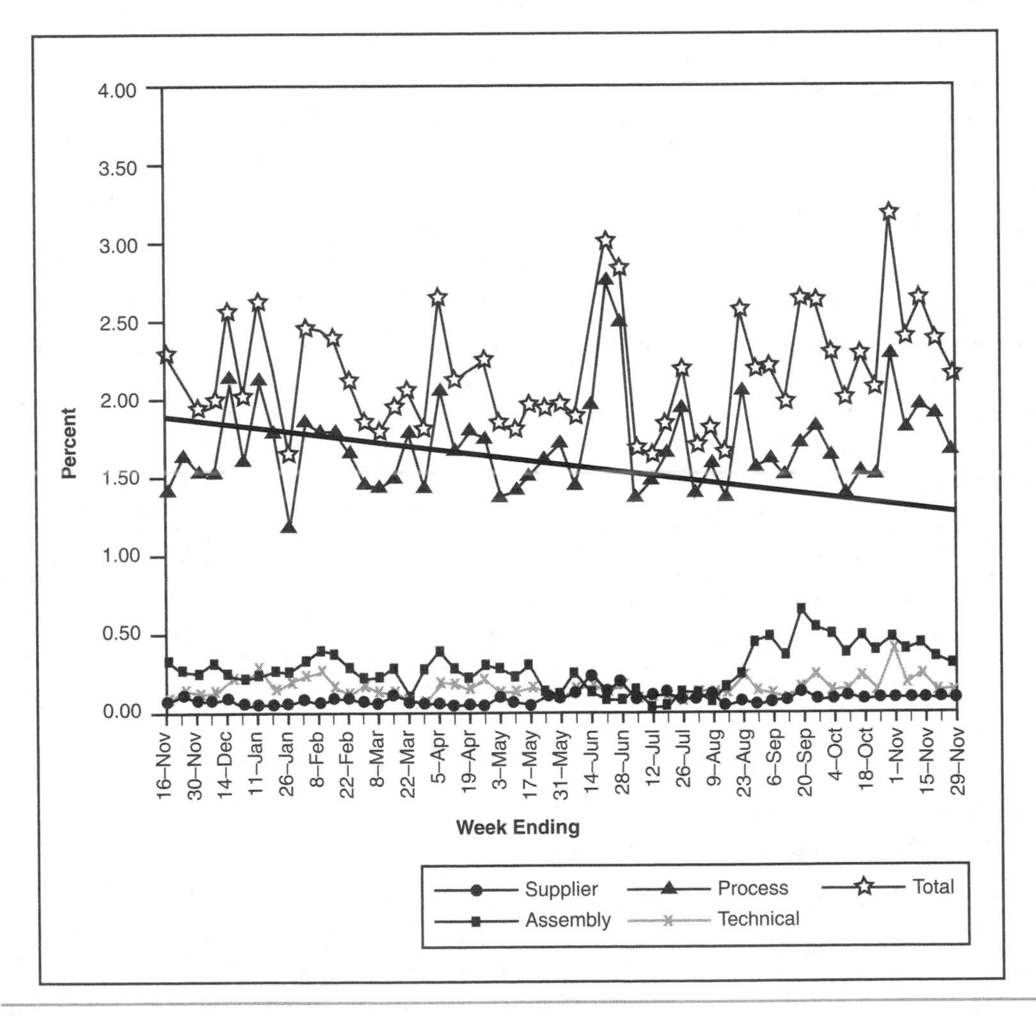

Figure 6.4 Example trend chart for scrap with a goal line.

Key Metric—Rework. Rework is typically easy to find in a manufacturing environment, but it is generally not tracked very well. The only detriment in measuring rework is that it takes an outside set of eyes to really see what is reworked because we become blind to rework when it is stable and has been there a while. Rework is also an issue in business processes, such as in redoing a contract because essential elements were omitted. Here are some examples of rework.

1. In Manufacturing:
   a. Readjustment of subassembly
   b. Adjust-at-test items
   c. Reformulation of chemical batches
   d. Reprocess for contaminant removal
2. In Business Processes:
   a. Reapply for mortgage
   b. Fixing wrong restaurant bills
   c. Food sent back at restaurant for additional cooking/preparation
   d. Rescheduled airline flights and crews
3. In New Product Introductions:
   a. Engineering change notices
   b. Repricing after launch (sales promotions)
   c. Field repair tickets

TOP-LINE EXPECTATIONS—EFFECTIVENESS AND VALUE PROPOSITION

We're in the business of producing product and services that are the most efficient and effective, and those with the highest value proposition generally get rewarded in the marketplace. The impacts are two: market dominance and command of a higher price.

Top-line expectations are the most difficult to meet. These expectations capitalize on the linkages between external realities of the market and internal activities to remedy market gaps. How do you know where you stand with regard to effectiveness in closing the gaps and your current value? Look for metrics like these:

Strategic Thrusts and Key Metrics:

1. Revenue:
   a. Top-line sales growth

2. Percent Revenue from New Products:
    a. On market for three years or less
3. Market Share:
    a. Percent core market ownership
4. New Market Development:
    a. Entry into existing markets
    b. Market capture of new markets
5. New Customers:
    a. Percent who used to purchase from competitors
    b. Percent who never owned/purchased before
6. Percent Repeat Customers
7. Percent Lost Customers

SIX SIGMA PARTICIPATION EXPECTATIONS

Participation expectations are determined within the goal trees. All divisions and functional groups will ultimately participate, and a phased participation is fully acceptable. Most Six Sigma deployments represent "All or some." Some business units may have priorities that will impede a new deployment, such as reorganization or SAP deployment, or perhaps they have just been merged with your organization and other changes are needed prior to deploying Six Sigma. In your program plan, you may want to have two or three phases to communicate with the organization when certain businesses or functional groups begin deployment.

1. Percent Business Units Participation:
    a. All business units should be deploying Six Sigma.
2. Percent Functional Groups Participation:
    a. All functional groups should be participating in Six Sigma.
3. Black Belt/Green Belt Percent Time Dedicated:
    a. 100 percent for Black Belts.
    b. 50 percent for Green Belts.
4. Black Belt/Green Belt Weekly Champion Meetings:
    a. Every Belt has weekly Champion meetings.

   b. Champions hold best-practice meetings quarterly or monthly at the beginning.
5. Black Belt/Green Belt # Sponsor Meetings:
   a. Monthly.
   b. Quarterly.
6. Number of Champion Project Reviews.
7. Percent of Annual Training Plan Complete.
8. Leadership Kickoff/Close at Training:
   a. Management level.
   b. Percent participating.

Measuring participation is not easy. But setting aggressive financial targets and other metrics will tell you whether everyone is on board or not. Nothing surpasses the energy as being there and asking the right questions. Several CEOs I know (Larry Bossidy, Fred Poses, George Fisher, and Paul Norris) were known for their extensive travel throughout their company and arriving at each destination with good questions about the Six Sigma deployment.

Selecting the right metrics—financial and performance—and expecting the right participation is the foundation upon which you will build your Six Sigma deployment. Get these two components defined before you launch your deployment, ensuring that all are talking off the same script.

PART II
THE FIRST 90 DAYS

7 Defining the Six Sigma Project Scope

With Dan Kutz

We now commence the first 90 days of your Six Sigma deployment. A sizable front-end portion consists of the process of planning the deployment. When doing a quick review of the activity that must be initiated, one can clearly understand the important of good upfront thinking and planning. The following list shows at least 14 different milestones that must be addressed during the initial stages of the deployment. Determining the scope of the deployment generates the milestones that must be accomplished for the deployment. There is a huge difference between the milestones for a small pilot deployment versus a large corporate-wide deployment.

1. Establish the Six Sigma leadership team.
2. Identify Corporate Critical Ys (metrics) and set goals.
3. Identify the timeline for deploying.
4. Black Belts: Operations, Services, Product Development.
5. Green Belts: Operations, Services, Product Development.
6. Implement project tracking system.
7. Identify potential projects linked to Critical Ys.
8. Identify Champions and Black Belt candidates.
9. Prioritize the projects and assign to the appropriate Black Belts(s).

10. Complete a charter for each project, including financial impact.
11. Define the project review methods and frequency for your business and plant or area.
12. Resolve if the resources and leadership commitment are in place to make the project(s) successful.
13. Determine the communications methods for your plant and business.
14. Develop reward and recognition guidelines.

The first element of the deployment is the scope of the Six Sigma deployment over the next 12 months. The first milestone of the deployment philosophy is to establish the business case first. Six Sigma—especially when including Lean Enterprise—includes many different business applications. The three usual applications include (1) manufacturing operations; (2) business process support; and (3) new product development.

That does not include other areas like marketing, sales, supply chain management, safety, and enterprise systems. So, deciding what the scope of the Six Sigma deployment is and where the focus will be is critical at the onset of the deployment. Therefore, different processes use different forms of Six Sigma and their outcomes:

- Transactional Six Sigma for business processes: Optimizing process flow and accuracy.
- Operational Six Sigma: Optimizing product and process flow and accuracy.
- Lean Sigma: Products and special tools for process flow.
- DFSS (Design for Six Sigma): New products, operations, and market windows.

Six Sigma can easily move from a focus on one function (e.g., manufacturing) or to a comprehensive business-wide focus to include all functions. Also important are deciding whether the effort is a pilot effort, what to do about overseas operations, whether to do part of the company or the whole company, or how to interface Six Sigma with preexisting or ongoing initiatives.

CONNECTING SIX SIGMA TO OTHER INITIATIVES

Connecting Six Sigma with Earlier Initiatives. Every company has a history of launching numerous initiatives—some successful, some not. Every company has established great programs in the past. The goal is to dovetail Six Sigma with the past good things and not bill Six Sigma as a replacement to those past accomplishments. For example, a company has implemented a good TQM program in the past; this company can pitch Six Sigma as the natural extension of the TQM program that adds advanced tools and methods to the effort. This approach acknowledges the good work done by the company in the past.

To effectively link Six Sigma to other initiatives, there might be a requirement to customize the Six Sigma approach to incorporate the tools and terminology of past successful programs. For example, AlliedSignal under Larry Bossidy had successfully launched Total Quality Leadership (TQL) and Total Quality for Speed (TQS) before it launched Six Sigma.

To allow those two initiatives to support Six Sigma, AlliedSignal chose to name their Six Sigma program Operational Excellence. During the initial pitches, great care was taken to demonstrate how Operational Excellence would build on the tools and methods learned in TQL and TQM.

3M had also done a lot of work in quality before their Six Sigma launch. There was great care to establish customization teams from 3M to work with the external consultants in customizing the 3M Six Sigma approach. This way, there was a direct link between Six Sigma and previous initiatives.

Effectively dovetailing your Six Sigma program with prior initiatives sends the message to your company that you are trying to extend and optimize the good work of the past. If you disregard past initiatives, Six Sigma might look like it's replacing a lot of things that were working fine already. That can lead to frustration and confusion among your troops. Six Sigma will be placed at the risk of becoming the infamous program of the month. Also, effective dovetailing helps the impending change not look quite so radical, because it simply extends old ideas.

One company we worked with had a marginal TQM program that they were never really happy with. They dovetailed Six Sigma into that by saying that Six Sigma was putting the brains back into TQM. Harsh, granted, but they were acknowledging that the TQM program left a lot to be desired, and that a lot of the concepts were important enough to get more advanced methodologies.

Connecting Six Sigma to Current Initiatives. When launching Six Sigma while other initiatives are currently underway demands a need for initiative coordination. For example, we had one client launching an SAP deployment at the same time as a Lean-Six Sigma deployment. They folded Lean techniques directly into the SAP deployment. That made sense because SAP automates business processes and Lean streamlines business processes. To streamline and simplify business processes before automating them was logical. I have a mantra for business process improvement: Simply, Standardize, Automate.

The other option is to make the choice to delay Six Sigma until the other initiatives are completed. We've seen potential clients do that, especially if they are in the middle or just beginning an SAP deployment. Because all initiatives require resources to be successful, the decision should focus on whether you can resource the initiatives you have on the table.

If the decision is to deploy Six Sigma while other initiatives are in progress, the communication of this should acknowledge that numerous initiatives are being executed and explain how these initiatives relate to the company's vision and mission. All these initiatives are essential to the future success of the company.

I watched Larry Bossidy confront this issue with some of his leaders. There were several initiatives in progress while Six Sigma was deployed—there was a subtle cry of "initiative overload." When asked about the overload, Larry asked, "If our vision is to be a premier company, which initiative should we stop?" The silence was deafening, and the point was made. To be excellent requires a lot of hard work from a lot of people.

PILOT OR FULL DEPLOYMENT?

There seems to be a common question before a new initiative: "Should we do it corporate wide or on a small scale (pilot)?" Either approach is correct, depending on the company. Strong corporate leaders such as Jack Welch, Larry Bossidy, Ed Breen, and Jim McNerney each launched Six Sigma across the corporation.

The concept of a small pilot didn't even enter their minds. However, numerous companies—such as 3M, Rohme and Haas, Johnson & Johnson, and Tyco—did start with pilot projects, which ultimately lead to corporate-wide programs.

Pilot Programs. Pilot programs serve the purpose of proving the concept. The advantage is that if the concept is wrong, the damage is contained in a small area of the company. However, if the concept is right, then the whole company has lost a significant amount of time to deployment and results. Programs of the month often start off as pilots, so you have that stigma to compete against. Also, pilots can be construed as leadership hesitancy for the program.

As you probably gathered, I'm not always in favor of pilots. But I have seen pilots work and be the right thing for some companies to do. The chief risk is that Six Sigma is launched by training a bunch of Black Belts, but their projects are poorly defined, the right resources are not available, and no one is truly accountable for the success or failure of the pilot. Pending a pilot Six Sigma launch, the pilot must be launched as a mini-deployment to include the same milestones as a corporate-wide deployment.

These are the 14 milestones presented at the beginning of this chapter. Without providing the Six Sigma pilot effort full support, you are dooming the pilot to failure. In addition, the infrastructure and methods developed in the pilot will probably be used in the corporate deployment. Customization of the Six Sigma training material will be much easier because you can use the data from the pilot to support the Six Sigma concepts.

Full Deployments. Full corporate-wide deployments have advantages that are apparent if the company's leadership takes the deployment seriously. However, the risk of a full deployment is that full deployments take huge top leadership commitment, which translates to a lot of time. The prelaunch deployment plan is required to be complete, detailed, and well disseminated. The training requirement, even for the first 90 days, is extensive. To add to the complexity, a full deployment is multifunctional. And finally, new positions created from the necessity to drive the program must be staffed.

The advantages of a full deployment are many. First, because the business case should drive the scope of the deployment, establishing the scope of the Six Sigma program to address the gaps in the external realities is possible. Company leaders are involved in the decisions concerning accountability and expectations at the onset. And, a company can expect a great ROI within six–eight months. The corporate culture is positively changed because of results, recognition, and rewards. You will have followed Kotter's eight steps (see Chapter 3, "Six Sigma Launch Philosophy," for detail) and reaped the benefits quickly.

Full deployments rely on the organization knowing what's going on. The corporate leadership has the window to systematically link their strategy to the annual operating plan and to Six Sigma projects simultaneously. So, the vision of closing gaps is now possible with a full deployment.

The risks of the full deployment are small. The first risk is that not all of the senior leaders will be on board for the deployment. CEOs like Welch and Jim McNerney resolved that by not letting Six Sigma be an option. But not all CEOs are willing to take that stance. The deployment will require vast amounts of time, so the risk is, Will your leaders be willing to spend the time? Rigorous and systematic program reviews during frequent visits around the company will tell the tale of commitment.

The history of Six Sigma shows that full deployments achieve results in order of magnitude well above pilot programs. If you want to change the culture, change the culture—it's as simple as that!

By Business Unit or Division?

Although full deployments are recommended, the reality is that full deployments are never completely full. Generally, most of the company is participating in Six Sigma, but not all of the company is taking part. For example, a new acquisition may not be ready to participate because they are still trying to transition from the old way of doing things to the new way of doing things.

So the question is, Do you select certain business units to start or entire divisions to start? The advantage of starting with some, but not all, of the divisions is that you gain some of the advantages of a full deployment and some of the advantages of a pilot deployment simultaneously.

Launching a Six Sigma deployment by division is probably less risky than launching by business unit. Because a division's organization and behavior are close to that of the entire corporation, each division represents a full deployment across all its business units. Each division creates the infrastructure it needs to drive the program forward.

If the launch is business unit by business unit, each business unit becomes its own micro corporation. The deployment roadmaps, the support infrastructure, the priorities, and the expectations will be unique from business unit to business unit. It will be quite a mess to try to standardize across divisions after the business units have gotten started. Going business unit by business unit will also end up costing more money in consulting because of numerous individually defined programs.

The wildly successful deployments of Six Sigma have had a small but centralized organizational infrastructure. For example, I lead the Six Sigma effort in the $4 billion engineered materials sector of the old AlliedSignal. I was definitely a sector HQ resource, and my staff only consisted of three or four Master Black Belts and some administrative assistance.

All other resources were decentralized and belonged to the divisions and business units. The advantage of the centralized office was the highly efficient coordination of the effort. We would ensure that all divisions and business units used the same project tracking system. We organized the annual training plan for Black Belts to foster efficiency and cost effectiveness. We standardized certification requirements for Master Black Belts, Black Belts, and Green Belts and maintained a sector register of those people who were certified.

In any large company, there will exist some business units that just don't want to play. These are usually the business units that are doing well and really don't think they need the help. I had one of those business units in my sector that just didn't play. I was upset about it, but my boss, Donnee Ramelli (who is now President of GM University), told me not to worry about it. That particular business unit accounted for less than 5 percent of the sector's revenue, and all the other business units were playing hardball. So, make sure that the divisions and business units that account for the highest business impact on the company are in the game. The others will come on board soon enough.

DEPLOYMENT DOMESTICALLY OR GLOBALLY?

Global companies always have to face the question of when to deploy Six Sigma outside their domestic boundaries. This was especially true for Honeywell, GE, 3M, and Tyco. To drive a successful global Six Sigma deployment is wrought with issues such as culture differences, language differences, and economic differences, to say nothing about logistic difficulties. The choice, of course, is whether to launch globally simultaneously or to use a phased approach to address the global nature of the company.

Considering the complexity of the global environment, it makes sense to get your ducks in a row domestically before going global. That way, the experience gained can be readily leveraged in the global arena. You have experienced many of the failure modes and worked through them. Your training material has been customized and is mature. And, you have a critical mass of experienced people who can aid in the global deployment.

However, speed is of the essence as you perform a phased deployment plan. Your global units may feel slighted if they have to wait too long for Six Sigma. It's best to bring the global players on board within months after the domestic launch. Also, you can involve global players right from the beginning by sending them to domestic training events. I had several students from Europe and Asia in my U.S. classes, and it worked out great.

Leadership training can be initiated as soon as the training material has been translated. You can give aggressive global leaders a head start and support this way. You really have to dovetail the deployment globally the way the culture will accept.

The global plan is documented in the overall Six Sigma deployment plan so everyone knows what is going to happen. The longer you wait to go global, the more you miss out on business results. I saw a German Black Belt improve capacity by 50 percent in a sold-out market in four months. That yielded millions of dollars for the corporation.

SIX SIGMA PROGRAM CONTENT

Because Six Sigma has such a wide range of applications, the focus of the program for the first three years should be established. Many companies focus the first year on manufacturing, the second on product design, and the third on business processes. Again, the program content should directly address your external realities.

Integrating Lean Enterprise. Recently, I am seeing more and more companies integrating Lean methods with Six Sigma methods. This, of course, is a wise thing to do because Lean focuses on speed and flow and Six Sigma focuses on accuracy and removing variability. Both efforts work well hand in hand. I have seen the failure mode of a company trying to launch Lean and Six Sigma as separate and parallel initiatives. The result has often been the creation of two competing camps, with each camp trying to lay claim to the other camp's results.

It turns out that the companies that launched the two methods as Lean Sigma have been successful. In fact, for some companies, it is best to start with Lean training, get quick results, foster enthusiasm, and then bring in Six Sigma. A company can do quite a few Lean Kaizen events based on a manageable level of training. The team aspect of Lean carries over directly into Six Sigma projects.

Because the focus of both programs is process improvement, they work very well in ensemble. Six Sigma offers tools and approaches that Lean doesn't, and Lean offers tools and approaches that Six Sigma doesn't. By combining the two initiatives into Lean Sigma, you end up deploying a wide variety of tools to a wide variety of people. These people have learned to use the right tool in the right place. There are no competing camps, just the competition of the market place. Fast, dramatic results rule each day.

The last way to incorporate both methods is to launch Lean and Six Sigma one at a time. By doing that, you tend to lose the integration of the two, and you tend toward different infrastructures to drive each one. By launching both at the same time, you can launch off the same infrastructure and the same tracking methodology. These grand efficiencies rule the day.

Other Six Sigma Programs. There are an exceptional number of Six Sigma programs, of which all or some may be launched. Figure 7.1 shows a high-level description of where Six Sigma plays in the process of identifying value, delivering value, and communicating value. The most popular program is Six Sigma for Operations. This program is associated with the birth of Six Sigma at Motorola. This program is aimed directly toward manufacturing operations. For manufacturing companies, this will always be the program to be included in the first year of deployment.

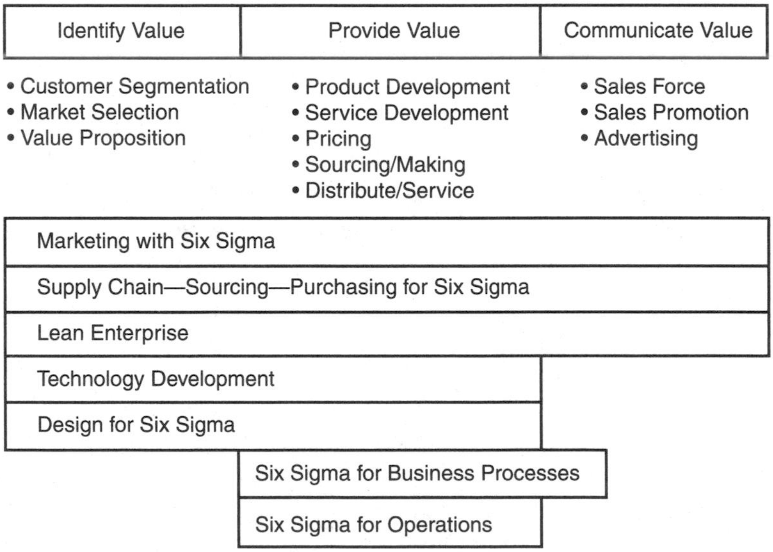

Identify Value	Provide Value	Communicate Value
• Customer Segmentation • Market Selection • Value Proposition	• Product Development • Service Development • Pricing • Sourcing/Making • Distribute/Service	• Sales Force • Sales Promotion • Advertising

Marketing with Six Sigma
Supply Chain—Sourcing—Purchasing for Six Sigma
Lean Enterprise
Technology Development
Design for Six Sigma
Six Sigma for Business Processes
Six Sigma for Operations

Figure 7.1 Six Sigma programs, as related to identifying, providing, and communicating value.

The MAIC (Measure, Analyze, Improve, Control) roadmap is the heart of the program. The operations program is driven by metrics that represent quality and cost improvement, as well as improvements in speed, capacity, and productivity. The Operational Six Sigma program can include many courses such as supply chain management, equipment reliability, process measurement systems, and Lean methodologies.

The next popular program is Design for Six Sigma (DFSS). While the Operational Six Sigma drives bottom-line growth, DFSS drives top-line growth through the efficient development of new products and services. Although the sense of DFSS is that it is aimed at technical resources, a comprehensive DFSS also includes sales, marketing, and technical development. A large part of DFSS includes programs addressing the voice of the customer along with the technical product development tools that are necessary to create and launch a new product into the marketplace.

A relatively recent program in Six Sigma is Transactional Six Sigma. This very important program addresses business process improvement or business process development. Software development is also included in this area. Because transactional processes are very different from manufacturing processes, a different roadmap and set of tools are needed.

With Transactional Six Sigma, Six Sigma can literally be driven into all parts of the company. For example, just the legal department in a corporation has control of over 45 different processes. Just think of the number of business processes you manage (or try to manage) across your company. There are probably 2,000 or more processes your people manage.

To get Six Sigma started on the right foot, a series of assessments are recommended. There are several assessment areas to include, such as operations, technology, and business processes. These assessments are valuable because they give a database's status on what's working and what's not. Assessments are a great way to establish strengths and weaknesses and to develop initial Six Sigma project lists.

SCHEDULING EVENTS

With any Six Sigma deployment and especially a deployment that is widely scoped, there are a very large number of events that are to be scheduled. This is one of the major functions of a centralized Six Sigma group within a company. The following table gives a partial Microsoft Project output for an early Six Sigma deployment. Listed are the events and the number of days necessary. The actual file includes start dates and resources. There are several scheduling and logistics pitfalls:

1. Notifying the participants too late.
2. Not considering holiday schedules.
3. Not planning for transfer of responsibilities.

The requirement to meet before deploying Six Sigma is to understand the scope of the deployment. Considering the options and how those options relate to the company and its external realities well in advance of the kickoff will help solidify and clarify Six Sigma and how it fits into your big picture. Six Sigma is complex but can be made simple with the right upfront considerations. Table 7.1 gives you some idea of the number of events and timing to get to your first extensive round of Six Sigma training within the target 90 days.

Table 7.1 Deployment Milestones for the First 90 days of the Six Sigma Deployment

Day	Event	Milestone
1	Executive Interviews/Alignment	Meet with corporate leadership teams • Review current state (ID and rank evidence/impact) • Discuss future state (measures of success and benefits of it) • Identify expectations (timing; division of labor) • Strategy and business plan review • Identify issues and concerns • Align program

7	Assessments	Conduct a Series of 3–5 Day Business Assessments
		• Core processes
		• Production processes
		    • Six Sigma assessment
		    • Lean assessment
		• Stage gate processes (for DFSS programs)
		• Candidate capability to align skills and gaps, if existing GB or BB candidates are selected for further training (BB, MFSS, DFSS, Transactional GB/BB)

14	Executive Workshop	Introduce Program to Company Executives • Align leadership support with program • Develop project clusters and goal trees • ID roles and responsibilities • Champion selection

Day	Event	Milestone
21	Strategy Workshop	If Required to Develop or Reenergize Corporate Strategy • Horizon planning • Business metrics • Gap analysis
28	Finance Workshop	Align Financial Team • Identify business drivers • Link program with financial performance, metrics, and systems • Establish guidelines and metrics • Define financial team's role
31	Human Resources Workshop	Align Human Resources Team • Succession planning and career planning • Measuring Belt performance • Reward and recognition
42	Champion Workshop	Align Champions (Mid- and High-Level Program Directors and Mentors) • Introduce program • Identify organization's needs • Project identification, prioritization, and selection • Roles and responsibilities • Metric definition and tracking systems • Project review planning • Belt selection • Project definition and chartering
56	Program Planning	Align the Program to Meet Business Needs • Training module selection and customization • Program manager and instructor selection • Belt selection review • Project charter review
90	BB/GB or VOC Wave Week 1	Initiate Six Sigma Training Program

8 Defining the Six Sigma Infrastructure

With Dan Kutz

Any major change initiative requires a clearly defined supporting infrastructure to drive the program. Infrastructure is defined as the underlying foundation and basic framework of personnel and supporting systems needed to support Six Sigma deployment activities. Because every part of a company participates in Six Sigma activities, the infrastructure must be clear, consistent, and comprehensive.

An effective infrastructure facilitates the development of the core competency that will establish and link Six Sigma project teams to (1) projects, (2) financial targets, and (3) the strategic plan. These project teams will be multifunctional and will need multifunctional support to execute the projects.

If Six Sigma has any chance of being successful, the infrastructure will span from the CEO and his leadership team to business leaders and to people executing the projects. Remember we learned earlier that one of Kotter's eight stages of leader change is "Create a Guiding Coalition." Thus, there is the goal of the Six Sigma infrastructure.

The infrastructure creates a strong network among the Executive Team, the Six Sigma Champions, the Belts, and the functions and businesses. This makes sense because the CEO's leadership team holds the accountability for executing the corporate strategic plan, and Six Sigma projects are instrumental in moving along the strategic plan.

One learning challenge of a Six Sigma deployment involves training the Six Sigma project teams. The human resources on these teams must learn how to work as a Six Sigma team. A new roadmap and a new set of tools, plus a more distinct focus on project accountability, add to the changes confronted by an organization when creating a Six Sigma environment.

Equally more important and complex is the learning challenge of the senior executives. Teaching the leadership team to learn how to lead a team-based organization is essential to strategic and long-term success. Because executing the strategy is a clear responsibility to which the senior executives are accountable, it follows that becoming a dynamic team leader within the Six Sigma deployment will support the strategic efforts.

Executing a good strategic plan entails the coordination of multifunctional internal activities. Senior executives must learn to deal with a multifunctional arena rather than the traditional functions. Hundreds of Six Sigma teams launched simultaneously is the outcome of an exemplary deployment of Six Sigma. Each of these teams need at minimum

1. Clear purpose for the Six Sigma team structure.
2. Clear Six Sigma program expectations.
3. Six Sigma project charters.
4. Six Sigma infrastructure tracking the number of teams.
5. Centralized repository for project results.
6. Six Sigma team goals.
7. Six Sigma team reporting mechanism.
8. Rewards and recognition alignment.
9. Six Sigma training and development plan.
10. Six Sigma team performance measures.
11. Deployment management of Six Sigma teams.

To accomplish all of the preceding requirements demands an extensive infrastructure with supporting systems. Preexisting resources are largely used to staff this infrastructure. Deploying a Six Sigma program, however, does not assume a requirement to add outside resources in a lot of new positions. The additional costs will usually have to do with the external consulting group you hire.

For example, the only resource that Larry Bossidy added when he launched Six Sigma into AlliedSignal was a corporate program leader. Larry brought in Richard Schroeder from ABB to drive the program. All the other resources for AlliedSignal's Six Sigma program already existed within the company. A small number of additional resources were added by the businesses as needed.

Because accountability represents the hallmark of successful Six Sigma deployments, defining the Six Sigma infrastructure and staffing and training the infrastructure players should happen very early in the Six Sigma deployment. Training is essential since, as Larry Bossidy has advised in his book, *Confronting Reality*, you must "Learn the guts of the initiative." He also adds that key members of the leadership team should learn the guts of the initiative. Early leadership training becomes a natural part of Six Sigma deployments to allow the program leaders to learn the guts of Six Sigma before the program gets too far along.

Defining the Six Sigma infrastructure is a little tricky. There should be a small centralized unit to ensure consistency and cost effectiveness of Six Sigma activities across the businesses and functions. There should also be a decentralized process that allows each business and function to tailor the Six Sigma deployment to its special needs. There is a big difference in deploying Six Sigma into the Human Resources function when compared to deploying into product development and R&D. So, our recommended infrastructure has both centralized and decentralized elements in it.

THE SIX SIGMA INFRASTRUCTURE

The aspects of the infrastructure that will be addressed are the following:

- CEO and Executive Team—Corporate and Businesses
- Six Sigma Initiative Champion—Corporate
- Six Sigma Deployment Champions—Business Unit
- Six Sigma Project Champions
- Master Black Belts
- Black Belts
- Green Belts
- Six Sigma Project Team Members
- Six Sigma Finance Support
- Six Sigma Human Resource Support
- Six Sigma Project Tracking
- Six Sigma Steering Teams
- Six Sigma Certifications

CEO AND EXECUTIVE TEAM

The commitment of the CEO and the Executive Team is essential to the success of the Six Sigma deployment. The Executive Team consists of the CEO and his or her first-level reports or sometimes his or her second-level reports. There is a difference between a show of commitment and a show of support. Larry Bossidy and Fred Poses constantly demonstrated commitment of the first years of the AlliedSignal deployment.

Larry met with every Black Belt class that was taught at the AlliedSignal Learning Center in Morristown, New Jersey. He held detailed program reviews every quarter with every segment in the company. Leaders that simply show support will not stop activity in the initiative, but will not actively drive the program in a strategic direction. Supporters tend not to create a sense of urgency around the initiative.

Larry says in *Confronting Reality*, "Leaders must have the courage of their convictions to follow through, and they have to be both inspiring and unrelenting. People need to know that there will be consequences for not getting behind the initiative." This group must be prepared to lead the Six Sigma as a strategic initiative, complete with strategic goals and expectations. Because Six Sigma programs historically generate revenue for the company based on savings, the Six Sigma deployment should be run as a new business within the company. My Black Belts generated a validated $95 million during the first year. This was roughly equivalent to the same results of adding a new business with the first year revenue of almost $1 billion.

The book titled *The Leadership Challenge* by Kouzes and Posner (1995) sets forth the fundamentals of leadership with the following five principles:

1. Challenge the process.
2. Inspire a shared vision.
3. Enable others to act.
4. Model the way.
5. Encourage the heart.

Some process has surely been challenged or you wouldn't be deploying Six Sigma in the first place. However, Six Sigma will be a new process within the company and must be challenged until the process becomes institutionalized and robust.

Once the Executive Team learns the guts of Six Sigma, the shared vision that is created should be dynamic and exciting. Because of the academic complexity of Six Sigma, the corporate-wide training plan will be developed. The training in the Six Sigma roadmaps and tools, along with the support and commitment to use them, will enable others to act.

Modeling the way means that the Executive Team has learned the language of Six Sigma, talking to the Black Belts, Green Belts, and Six Sigma Champions on every business trip. They hold detailed program reviews consistently and ask hard questions at those reviews. The Executive Team appears as if they have been trained as Black Belts themselves. Larry Bossidy and Fred Poses underwent extensive training in Six Sigma and knew it well. They were so inspiring that my Black Belts would follow those guys anywhere after they addressed a group of Black Belts.

Well-thought-out rewards and recognition systems encourage the heart of the resources committed to Six Sigma. The Executive Team makes sure that these new systems are in place. The Executive Team ensures

- Personal understanding and commitment to the deployment.
- Definition of the scope of the Six Sigma deployment.
- Integration of Six Sigma into other programs and initiatives.
- Strongly developed rationale for Six Sigma:
  - Ties to benefits everyone can understand.
  - Avoids the flavor of the month.
- Six Sigma program integrity, which includes people selection, credibility of results, institutionalization of methods, and reviews.
- The right leadership, ownership, and accountability in all Six Sigma roles at all levels.
- Promotion of Six Sigma across business units and functions.
- Business goals and metrics are understood and are the basis of all improvement efforts.
- Linkage of operational performance metrics (little Ys) to business metrics (Critical Ys).
- Personal engagement: Training class visits and rigorous program and project reviews.
- Communications of program expectations to Six Sigma Champions:
  - Support and mentor Champions.
  - Set accountability for Champions.
- Dedication of time for training and application.
- Questions about Six Sigma prepared for each site visit.

Some other roles and responsibilities for the Executive Team might include the following:

- Understanding methods and tools well enough to ask questions.
- Holding themselves and others accountable.
- Expecting and demanding the use of Six Sigma tools.
- Celebrating Six Sigma successes.

Leading a successful initiative is not easy. I saw Fred Poses at AlliedSignal spend 20–30 percent of his time for about three months to the Six Sigma launch. At the beginning, he held face-to-face meetings with all his General Managers, and they would explain what their projects would be, the forecasted financial impact of the projects, and who the Champions and Black Belts were going to be. He always asked, "Are these the right projects?" and "Are these the right people?" He also fired a plant manager who was subtly blocking Six Sigma. No one had any doubt about Fred's commitment. Other leaders demonstrated their commitment in other ways:

- **Motorola.** Bob Galvin schedules Six Sigma quality reports first on the agenda for the monthly operations meeting and then leaves afterward.
- **GE.** Jack Welch educates himself on Six Sigma with multiday training sessions, completes a Green Belt project, and makes Six Sigma a leadership requirement.
- **American Standard.** CEO Fred Poses briefs himself on key projects underway at a plant/site before visiting that site.
- **Cummins.** A Six Sigma Champion is included on the BU Leadership team, with progress reviewed at every monthly BU review. The Senior VP personally reviews every Black Belt project once per month.
- **Eaton Corporation.** The Division President authorizes ROI-justified headcount increase to backfill key roles "vacated" by full-time Six Sigma players.
- **Celanese.** President Dave Weidman says no capital requests will be approved unless evidence of significant Black Belt and Green Belt work has been completed. He brought Six Sigma into Celanese from his AlliedSignal experience.
- **3M.** The new CEO, Jim McNerney, launches the first Black Belt waves three weeks after the executive training workshop and attends the four-day executive kickoff meeting.

Early involvement in deployment planning is especially important to the executive team. Taking the right actions in the right order works to create a model Six Sigma deployment.

If the Executive Team is not fired up right from the beginning, the company will sense the lack of commitment and the launch will be considered another program of the month. A successful change initiative is the legacy that the leadership leaves the company. Six Sigma lived on for a long time after CEO, Bob Galvin, championed the effort in 1987.

CEO and Executive Team—Training. Because Six Sigma has a unique language and technology and because many important business decisions affect the deployment, a substantial amount of training before Six Sigma is launched is required. Chapter 10, "Creating Six Sigma Executive and Leadership Workshops," discusses the duration and the content extensively. Generally, during the first 90 days of the deployment, the executive training ranges from one day to four days. I recommend at least a two-day session because the training session is part education and part deployment workshop. Longer four-day sessions have worked well, leaving the leadership team with a clear concept of what Six Sigma is all about and a good start on a deployment plan with an action plan. A rough agenda would include such topics as

- Six Sigma overview
- Project selection
- People selection
- Tools and methodology
- High-level financial target development and first draft deployment plan
- Defined project clusters

Post-launch, the Executive Team might complete Green Belt training or an executive Green Belt training and might even complete a project. Jack Welch actually completed a Green Belt project on his watch.

SIX SIGMA INITIATIVE CHAMPION

The next role in the Six Sigma infrastructure is the Six Sigma Initiative Champion. This person is the full-time Six Sigma program leader who will ideally report directly to the CEO or a member of the Executive Team. He/she will lead Six Sigma as a strategic initiative across the corporation. This will be a centralized resource—for example, the CFO of a large railroad company was the Six Sigma leader.

The Initiative Champion has the responsibility for steering the initiative and will lead the Six Sigma Steering Team, which will have representation from across the business units. The Steering Team will include the Six Sigma Deployment Champions (described in the next section of this chapter) from the business units. The Initiative Champion will be charged to lead the overall deployment of Six Sigma, and be accountable for the strategic and financial impact of the program. Figure 8.1 shows a representation of the relationship between the Executive Team and the businesses (divisions). Each division's Deployment Champion reports (dotted line) to the Initiative Champion for the first Six Sigma Steering Team.

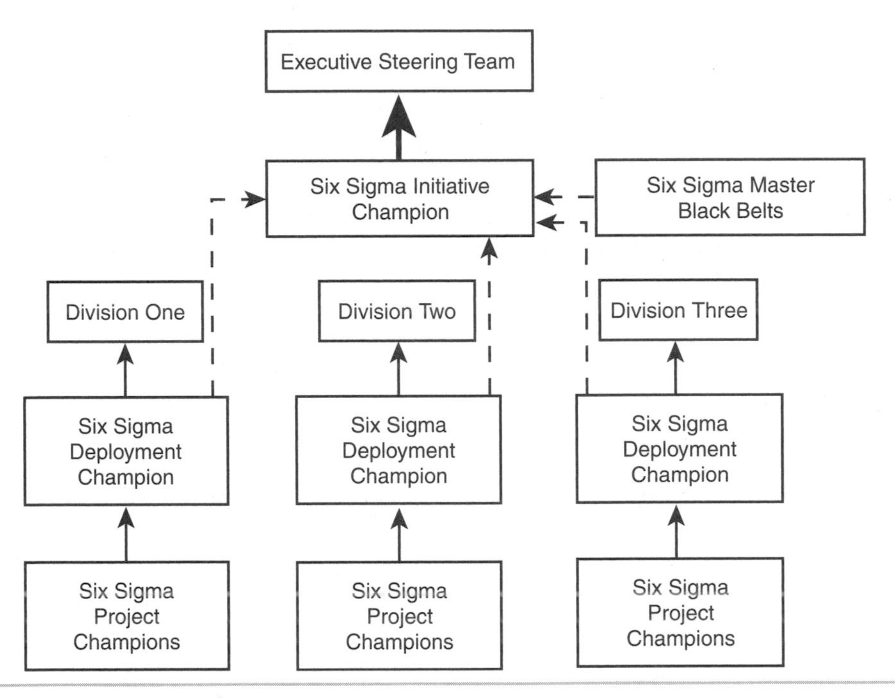

Figure 8.1 The Initiative Guiding Coalition. The Six Sigma Initiative Champion is a member of the Executive Steering Team and has a Deployment Champion in each business reporting as the Six Sigma Steering team.

In addition, the Initiative Champion will steer the evolution of Six Sigma over the years to reflect Six Sigma's alignment with the strategic plan. Don Linsenmann of DuPont has been doing this for several years. The Initiative Champion will also remove barriers for success and remove leaders who aren't fully supportive of the program. The Initiative Champion is the primary conduit between the Executive Team, business units, Six Sigma Champions, and other Six Sigma resources. The following is a bulleted list of the general responsibilities of the Initiative Champion:

- **Roles and Responsibilities**
  - Leads Six Sigma full time as a strategic initiative.
  - Responsible for the Overall Six Sigma steering team.
  - Promotes Six Sigma across the business.
  - Delivers program strategic and financial impact.
  - Ensures program integrity (people selection, credibility of results, integrity of methods, and reviews).
  - Controls the development of training curricula and training plans.
  - Removes barriers.
  - Involved in strategic and business direction planning.
  - Identifies cross-business opportunities and best practices.
  - Creates Six Sigma Steering Teams where needed.
  - Identifies leaders who divert resources from the effort, or who do not contribute.
  - Attends selected classes and project reviews (stays engaged).
  - Communicates expectations to Champions and Belts.
  - Ensures active involvement and commitment of senior executives.
  - Leads the evolution process for Six Sigma.

Because training is such a large part of a Six Sigma deployment, a priority for the Initiative Champion is to guide and coordinate the schedule of training events and the development of the training curricula. The Six Sigma training will be more cost effective and deployment effective if coordinated centrally by the Initiative Champion.

Selecting the Initiative Champion is the first show of commitment by the Executive Team. Who the team selects will directly reflect the importance of the Six Sigma program. I recommend the Initiative Champion to be a first-level report to CEO or some other member of the Executive Team (CFO, VP of Operations, VP of Technology, etc.). This person should be a seasoned executive with a strong history of driving results and a respected leader in the corporation. The CEO should select the Initiative Champion.

- **Training Requirements**
  - **In The First 90 Days:**
    - One-day executive session.
    - Kick-off and participate in all Champion sessions.
    - Kick-off and close Black Belt training.

- **Post-Launch:**
  - Partners with a Black Belt to complete a Green Belt-level project they select.
  - Black Belt or Green Belt training.

Six Sigma Initiative Champions have been defined differently for different companies. Many companies assign the Quality Officer to the job. That seems to make sense, but the program will seem like a quality program rather than a strategic business program. Six Sigma addresses much more than process quality, although Quality Officers have done a great job. In AlliedSignal, the Quality Officer, Jim Sierk, was assigned as the administrative head of the program. But AlliedSignal also hired Rich Schroeder, who had a very strong quality background, to run the program day to day and report to Jim.

The Cummins Engines CEO, Tim Solso, named one of his best operating officers, Frank McDonald, to be the program leader. Frank was well respected in the company and approached Six Sigma as an operational initiative. With Celanese, a large chemical company, the President, Dave Weidman, named one of his best operating officers, Jim Alder, to lead the program. Jim also treated Six Sigma as an operational excellence program and did a great job of ensuring that Six Sigma activities led to bottom-line results.

History shows that Initiative Champions with a deep business and operational background tend to drive the program strategically. It appears the best candidates are the ones who ultimately want to run a business when they complete their term as Initiative Champion. Jim McNerney at 3M assigned Brad Sauer to Six Sigma, and Brad now runs a 3M business. However, anyone who reports to the right person and has the respect of their peers will do a good job. The Initiative Champion should act as if he or she is the Chief Operating Officer and drive the program from that standpoint.

SIX SIGMA DEPLOYMENT CHAMPIONS—BUSINESS UNIT

For a large company with multiple large divisions, the deployment must take a decentralized focus. For example, in 1994, AlliedSignal had three major segments (business units): aerospace segment, automotive segment, and engineered materials segment. Each one of these segments had radically different products and markets, so it made sense that Six Sigma would be somewhat customized for each segment. GE is another classic example of a wide range of businesses. The GE Capital's Six Sigma deployment was radically different than the manufacturing divisions because their business was so different.

Each major business unit should have a Six Sigma Deployment Champion who leads the business program. These resources should report to the business leader or business operations officer but have dotted-line responsibilities to the corporate Six Sigma Initiative Leader to retain the consistency in program across the company. The

Deployment Champions will make up the Six Sigma steering lead by the Initiative Champion. As with the Six Sigma Initiative Champion, the Deployment Champions have a significant role within the business. Some of their roles and responsibilities include the following:

- **Roles and Responsibilities**
  - Lead Division or Functional Area Six Sigma efforts.
  - Lead Business Team workshops (with Master Black Belt support) to facilitate project identification, prioritization, and selection.
  - Manage project sets within their business area.
  - Coordinate with other Deployment Champions in other business units/areas.
  - Identify cross–business unit opportunities.
  - Report to Business Unit leadership and dotted line to the Initiative Champion.
  - Maintain current project portfolios and performance data.
  - Identify and communicate success and problems in delivering project results.
- **Who Are They?**
  - In a leadership position.
  - Involved in budget and strategic planning for the business/facility.
  - Has ability to remove roadblocks such as personnel, capital, and time constraints.
- **Selection**
  - Selected by Initiative Champion and Business Unit Leader.
  - CEO approves the selection.
- **Training Requirements**
  - **In The First 90 Days:**
    - One-day executive session.
    - Two-day Champion training.
    - Co-facilitate Business Team Workshops to facilitate project selection in their area.
  - **Post-Launch:**
    - Green Belt training and projects.

The Six Sigma Deployment Champions address the unique needs of their businesses and business strategies. They account for the business and financial results of Six Sigma within their businesses. They also work in concert as members of both the corporate Six

Sigma Steering Team and their own business steering teams. Identifying and deploying Six Sigma best practices across the corporation is an important role.

SIX SIGMA PROJECT CHAMPIONS

The most critical role in a Six Sigma deployment is the Six Sigma Project Champion. The heart and soul of any Six Sigma program is the network of Six Sigma Project Champions. They are the catalysts for institutionalizing the grand change resulting from Six Sigma. These people are responsible for facilitating the selection of Six Sigma projects, fulfilling the Six Sigma training plan, and ensuring that each selected project is successful in meeting the financial and performance goals set. They also are instrumental in creating project charters that define each project. In addition, they ensure that the right resources are available for the projects.

Each Project Champion will usually be a process owner or a department head and lead and mentor anywhere from 2 to 20 Black Belts (most of which they have helped to select) and an assorted number of Green Belts. Within each business or function, the Project Champions will be members of the Six Sigma steering team for that business or the corporation.

- **Roles and Responsibilities**
  - Communicate the Six Sigma initiative and Six Sigma strategy.
  - Select projects and people.
  - Track the progress of Belts (weekly reviews).
  - Breakdown barriers for Belts.
  - Create supporting systems:
    - Project and results databases.
    - Networks and resources.
    - Incentive and reward systems.
  - Maintain performance databases.
  - Develop Master Black Belts.
- **Who Are They?**
  - Leader in the business.
  - Involved in budget and strategic planning for the business/facility.
  - Has ability to remove roadblocks such as personnel, capital, and time constraints.
  - Direct influence over resources in the process:

        - Money.
        - People.
        - Time.
    - Influential in the strategic plan for the business.
    - Involved in budget and annual operating plan development.
    - Possesses good mentoring skills.
    - Motivated change agent.
    - Respected within the business for driving positive change.
- **Selection**
    - By Deployment Champions and Business Unit or Area Leadership.
    - Initiative Champion approval of selection.
- **Training Requirements**
    - **In The First 90 Days:**
        - Two four-day Champion training.
            - Six Sigma overview.
            - Tools and methodology.
            - Project selection and prioritization.
            - Project chartering.
            - Project tracking and mentoring of belts.
        - Co-facilitate and attend initial Belt training.
    - **Post-Launch:**
        - Green Belt training and project encouraged.

The best Champions are those who revel in conquering organizational barriers and driving change. The network of Project Champions creates the guiding coalition necessary to drive change. From my AlliedSignal experience as a Deployment Champion, I had 12 Project Champions on my steering team for the $4 billion per year engineered materials sector.

These Champions were instrumental in finding great projects to work on, finding great Black Belts to train, and getting great results ($95 million the first year). We met frequently and defined and drove action items to ensure Six Sigma was successful. They probably spent a minimum of an hour per week with each Black Belt or Green Belt they mentored. They consolidated and reported the financial results for each active project.

They kept our Black Belt training sessions full of great people. Probably the most important action they took to institutionalize Six Sigma was to identify Master Black Belt candidates to develop into internal consultants and trainers. There was never any doubt in my mind about who really ran the program—the Champions. Said best, a senior Vice President with Cummins contended, "There are no unsuccessful Black Belts, just unsuccessful Champions." The idea here is that the Champion is accountable for the success for every Belt under his or her network.

MASTER BLACK BELTS

The role of the Master Black Belt (MBB) in the Six Sigma infrastructure is a fairly recent occurrence. The Master Black Belt has become a new corporate position, but only since about 1996 or 1997. The Master Black represents the technical glue that holds the Six Sigma methodology together. The MBB ensures that Six Sigma training is institutionalized and moves the company from being dependent on external consultants to the company being self-sustaining with its own internal consultants.

The risk with MBBs is that they are relegated to a training role instead of generating revenue through their mentoring of Black and Green Belts. The MBB is much more valuable working in the field with project support for the BBs and GBs. Training should only be a secondary role. MBBs have the ability to lead large multifunctional projects using a large number of Black Belts and Green Belts.

The return on investment is high when external consultants perform the training and the internal MBBs support the Six Sigma deployment. When your MBB is tied up for two weeks per month in training events and another week for training preparation, the time for value-added support is minimized. MBBs should be viewed as operational leaders driving strategic improvement activities.

The recommendation is to create a *training steering team* consisting of the MBBs and selected Champions. This team can internalize all the training and can schedule the training among MBBs. This way, each individual MBB spends a small amount of his or her time in training. It's also a great way to define and deploy technical best practices and prioritize training development activities.

Commonly, the usual tour for the MBBs is two years after certification. After that, they usually roll into a business leadership role, with the ultimate goal of running a business. The MBB program is often viewed as a leadership development program, so candidates for the MBB program should definitely be high-potential personnel.

In the early months of Six Sigma deployment, the MBB developmental training is generally done by external consulting groups. This development would only occur for candidates that have been certified as Black Belt and have completed two to three outstanding projects. After a critical mass of MBBs are certified, the MBB development actions are carried out by the MBB steering team. Even then, for some highly technical topics, external help may be enlisted.

The following are listings for roles and responsibilities, who they are, selection process, and training requirements.

- **Roles and Responsibilities**
  - Generates revenue: Coach and support Black Belts for results.
  - Develop and deliver Six Sigma training.
  - Assist in project identification.
  - Partner with Six Sigma Champions.
  - Identify and deploy best practices.
  - Partners with Six Sigma Champions.
- **Who Are They?**
  - Successful Black Belts.
  - Certified Black Belts.
  - Known for successful projects.
  - More senior employees.
  - Strong coaching/mentoring skills.
  - Deep business understanding for project identification and chartering.
  - Technically adept.
  - Comfortable training.
  - Proven mentoring skills.
  - Proficiency in the Six Sigma tools.
  - Ability to teach concepts effectively.
  - Documented and validated success with the methodology on at least two major projects.
  - Desires to pursue advanced statistical tool training.
  - Ability to apply tools and methods to areas outside of their current focus:
    - Example: Manufacturing person applying tools to business process.
  - Project Champion endorsement.
- **Selection**
  - Prerequisite: Has completed Black Belt training and at least two very successful Black Belt projects.

  - Strong candidate for future company leadership roles.
  - Nominated by Program and/or Project Champions.
  - Selected by Program Leader.
  - Approved by CEO.
- **Training Requirements**
  - Launch:
    - Complete Black Belt training.
    - Complete two or three successful projects.
    - Teach Green Belts.
  - Post Launch:
    - One week per month for five months (25 days), plus two elective weeks (10 days).
    - Technical skills (50 percent).
    - Leadership and change management skills (30 percent).
    - Teaching skills (20 percent):
      - Statistical and Lean tools.
      - Project management.
      - Change leadership.
      - Teamwork.
      - Control systems.

The preceding lists indicate just how important the MBBs are to the infrastructure. They are technical experts who know how to get results using sophisticated techniques. They are also the future leaders of the company. There will be trepidation in the MBB candidate when selected. The candidate will want to know how this is going to affect his or her career. Human Resources must provide a career path for MBBs to ensure that they know it's a good deal.

A tendency for companies just starting Six Sigma is to hire MBBs from other companies. The thought is that the Six Sigma deployment can be accelerated with these resources. I suggest it is much better to develop internal resources to fulfill that role. Your Six Sigma consulting group can easily provide you with the Master Black Belt services during the first year of your launch. The process of developing Master Black Belts internally sends an important message that Six Sigma provides another career path for future leaders. The drawback from hiring from outside the company is that there is inconsistency in the way companies develop and certify their MBBs.

The externally sourced MBBs may arrive at your company a little light on technical skills and leadership skills. Companies and external consultants have different developmental roadmaps. My company recommends a minimum of five weeks of training plus electives. Some companies only require two weeks of training. These external MBBs will be constrained by the Six Sigma methods they learned at their previous company, which will be different from the Six Sigma you are deploying. They will be in conflict with the external consultants you have enlisted to launch the program. Instead of removing variability, these external MBBs will tend to introduce variability. Beside, it only takes about one year to develop an MBB internally. Why not invest in your own people to move Six Sigma forward?

SIX SIGMA BLACK BELTS

In continuing our discussion of "Belts," we will now turn our attention to Black Belts. Black Belts represent the coalition of complex problem solvers. They will become your process improvement experts. They are trained in detailed roadmaps for solving problems in different arenas: manufacturing, product development, transactional business processes, supply chain, health care, and others. The first candidates should be selected from among the highest potential resources:

- Future leaders
- Technical experts
- Great mentors and coaches

Like the selections of the Six Sigma Initiative Champion and Six Sigma Deployment Champion, the quality of the candidates selected in the Black Belt training launch will indicate the company's commitment to really making a difference. There is a tendency by the businesses in the early days of Six Sigma deployment to sandbag the student count with a number of mediocre candidates.

The thinking is, if this is just another program of the month, why should I send one of my high-potential people and lose them for four or five weeks of training? The candidates for Black Belt training should be carefully reviewed for each wave of training to ensure the right people are in the right places. However, the good news is that Black Belt training will convert a former "dirt bag" into a high-performing asset.

The company should target 75–100 percent time dedication for Black Belts. The level of time dedication is always a question at the beginning of the Six Sigma launch. The fear is that if the people are full time, how do you fill gaps in performance? I recommend that Black Belts are full time, but they really don't need to be.

The issue is that if you pick great, high-impact projects, the decision addressing the dedication time becomes easy. If your organization has identified a project it believes will deliver a million dollars (not that unusual for Six Sigma projects), then having the Black Belt dedicated only 50 percent of the time on the project is ridiculous.

I will say that pending the decision to create a bunch of 100 percent Black Belts, with the exception of the first project in which the Six Sigma roadmap is applied the first time, a 100 percent Black Belt should be driving at least two or three projects simultaneously. Whereas, a part-time Black Belt is only able to drive one project at a time. The cycle time for completing a series of projects in parallel fosters better ROIs than a series of projects done sequentially.

Because the average value of a well-defined Six Sigma Black Belt project is $250,000, you have the choice of levying the worth of each of your Black Belts of one million dollar plus for full-time resources or less than $750,000 for part-time resources. Either way, it's a great return on salary and benefits. It's also a great return on the four weeks of training—thus, the importance of having a dynamic Six Sigma supporting infrastructure to make sure all this happens.

The following lists provide you with a brief overview of a Black Belt's roles and responsibilities, their characteristics, selection process, and training requirements.

- **Roles and Responsibilities**
  - Lead strategic, high-impact process improvement projects.
  - Master basic and advanced quality tools and statistics.
  - Deploy techniques of measurement, analysis, improvement, and control.
  - Participate in intensive 4- to 5-week training program.
  - Manage a BB project as a major plant or system project.
  - Significant dedication of time to a project.
  - Full-time role for 1.5 to 3 years.
- **Who Are They?**
  - High potential as future leaders of groups or businesses (leadership qualities).
  - More senior employees.
  - Known and respected by area and business unit leaders.
  - Exposed to multiple departments and/or sites.
  - Employees known to "get important work done."
  - Technical experience in their area of current focus (either with the company or from recent employment).

    - Desire and ability to drive change.
    - Self-starter/self-directed.
    - Not satisfied with the status quo.
    - Mathematical competency.
    - High energy; driven.
    - Effective facilitator.
    - Works well in team environment.
- **Selection**
    - Deployment Champion, Project Champion, Business Unit Leader, and Department Heads nominate Black Belt candidates (after they have attended Champion training).
    - Key question: "Can this person deliver project results?"
    - Selection approved by Deployment Champion.
- **Training Requirements**
    - Intensive training in the shortest amount of time possible.
    - In-depth instruction on the tools in the improvement roadmap.
    - One week per month for four months (20 days).
    - Technical skills (70 percent).
    - Project leadership, presentation, and change management skills (30 percent).
    - One week per month for five months (25 days) for Design for Six Sigma Black Belts.

Training and developing Blacks Belts is a huge investment of money and time. Most companies certify about 2 percent of their population as Black Belts. This investment gets an outstanding return if there is an effective process to systematically prioritize and select projects, provide the resources necessary to complete the projects, and determine the final business impact of completed projects. The side effects of a Six Sigma deployment is that your organization will get better at these actions than they already are. These are the actions that will carry your company through a strategic plan and attain your final vision.

SIX SIGMA GREEN BELTS

Green Belts are the tactical arm of Six Sigma. The role of the Green Belt varies but is nonetheless important. They lead their own projects within their relatively narrow area. For example, an Iomega administrative assistant worked on a project to reduce and control travel expenses. Although this is not a major business process, it nonetheless accounted for a few million dollars in the G&A budget. She ended up saving the company about $250,000 a year in travel costs with a well-defined and controlled new process.

Green Belts may find themselves supporting a Black Belt in a large multifunctional project. A Black Belt working on implementing an online sales project may well have Green Belts in functions included in the project in support. This infrastructure allows Black Belts to pursue multiple large-scale projects because he or she has technical help for each project.

For example, while working in an ABB manufacturing plan in Athens, Georgia, I functioned as the plant Master Black Belt. We had no Black Belts, and I had to train my own Green Belts. I had about 10 Green Belts in a 300-person factory. In less than 10 months, we completed 30 projects and brought over one million dollars to the bottom line. All we would have needed to do that was one Black Belt and 10 Green Belts on the team—less than 4 percent of the population.

My own personal philosophy is that Six Sigma deployments are best when focused on training and deploying hundreds of Green Belts, especially if there is a core of Black Belts and Master Black Belts ready to support and mentor them. Jack Welch even made it a requirement that all his leaders were trained as Green Belts.

The tools set for Green Belts is very similar to that of the Black Belts minus the sophisticated statistics and experimental design tools. The Green Belt training is about 50 percent of the Black Belt training, so the ability to deploy a large number of Green Belts is enhanced.

Green Belts are especially effective while working in the business process area. The processes are usually more simple and the need for fancy statistical tools is less. The following lists provide you with a brief overview of a Green Belt's roles and responsibilities, their characteristics, selection process, and training requirements.

- **Roles and Responsibilities**
  - Participate in strategic, high-impact process improvement projects with Black Belts.
  - Lead high-impact projects within departments.
  - Master basic quality tools and statistics.
  - Help deploy techniques of measurement, analysis, improvement, and control.
  - Participate in a two-week training program.
  - Part-time role: 25 to 50 percent of time.
- **Who Are They?**
  - Lower level managers/staff.

  - Well respected within department.
  - Enthusiastic about process improvement.
  - Strong desire to learn.
- **Selection**
  - Deployment and Project Champions nominate Green Belt candidates and review volunteers who step forward.
  - Deployment Champion and Business Unit Leader approve selection.
  - Key question: "Can this person deliver project results in their area of work?"
- **Training Requirements**
  - Intensive training in the shortest amount of time possible.
  - In-depth instruction on the basic tools in the improvement roadmap.
  - One week per month for two months (8–10 days).
  - Technical skills (70 percent).
  - Project leadership, presentation, and change management skills (30 percent).
  - One week per month for three months (12–15 days) for Design for Six Sigma Green Belts.

A dynamic Green Belt program is important. Great Green Belts can go on to get certified as Black Belts or Master Black Belts. The Green Belt role allows a much wider variety of people to participate in Six Sigma and enjoy having a visual impact on the company. Each Green Belt can lead several teams per year, and your company can be characterized as a company that has hundreds of great teams working on process improvement all the time.

Green Belts add depth and critical mass to the deployment and institutionalization of Six Sigma. Most companies train 20 to 30 percent of their populations as Green Belts. They gain results through leading projects of a smaller scope, generally in their functional areas and supporting larger BB projects. Green Belts are worth about $100,000 per year if managed correctly.

Green Belts will transition to BBs with additional training, coaching, and expanded project focus. Finally, the GB program is a good training program for managers of Black Belts and Green Belts—for example, Test Lab Manager, Maintenance Manager, Customer Service Manager, and many others.

SIX SIGMA PROJECT TEAM MEMBERS

Every Six Sigma project has a team. The Black Belt is not the Lone Ranger. He or she also mobilizes a team, teaches and mentors them through a project, and strives for equal acknowledgment for results. The Six Sigma team is the most exciting part of Six Sigma. In Warren Bennis's book titled *Organizing Genius*, he contends that the great leaders of tomorrow exist in a fertile relationship with a Great Group.

The leaders that understand the importance of teamwork in today's complex marketplace will be the heroes. This follows the philosophical saying, "None of us is as smart as all of us." The concept of Great Groups is important, because every great leader has a Great Group in support. Six Sigma is an ideal way to create Great Groups.

Great Groups harness the creativity of every individual in the group. Just about every project review I've seen had documented a creative innovation that would not have been developed by an individual. The following lists will provide you with a brief overview of a project team member's roles and responsibilities, their characteristics, selection process, and training requirements.

- **Roles and Responsibilities**
  - Participate in strategic, high-impact process improvement projects.
  - Attend team meetings.
  - Apply basic quality tools within project.
  - Identify areas to apply techniques of measurement, analysis, improvement, and control in day-to-day job.
  - Communicate project progress to departments.
  - Part time role: 10–15 percent of time.
- **Who Are They?**
  - Level dependent upon project level.
  - Well respected within department.
  - Enthusiastic about process improvement.
  - Strong desire to learn and have an impact.
- **Selection**
  - Project Champions, Black Belts, and Green Belts.
  - Caution: Select team members who know the process area and who realistically can find time to contribute to the project.
- **Training Requirements**
  - Four to eight hours of process improvement roadmap and tools overview.

- Training covers *what* tools are and *when* they are used.
- Training may be delivered via e-Learning or by Belt, in advance of or at the start of project participation.

By giving Six Sigma Teams (Great Groups) a strategic, high-impact project, a process improvement expert (Black Belt or Green Belt), the resources to complete the project, and a clear problem-solving roadmap, the creativity soars! The Six Sigma tools are designed to consistently get the group to look at the problem in many different ways and, along the way, creative and innovative solutions are developed naturally.

SIX SIGMA FINANCE SUPPORT

Because the activities of the entire Six Sigma infrastructure are aimed at having a financial impact on the company, the finance team is an integral component to the infrastructure. The integration should happen early in the deployment. One of my very successful clients actually dedicated financial analysts solely to Six Sigma, creating Six Sigma Black Belts within the financial community. That is an approach to consider seriously.

- Establish common measures of project benefits.
- Sign off on project estimates and results.
- Provide input to the project selection and chartering process.
- Identify risks and opportunities associated with projects.
- Work with MBB, Champions, and Director to quantify benefits:
  - Confirm the Project Y objective as defined by the BB is appropriate and will result in "real" value.
- Participate in business-level Six Sigma reviews.
- Ensure Finance is an active participant in opportunity analysis with leadership team.
- Support identifying qualify and quantify opportunities.
- Determine potential impact to revenue growth, operating income, and cash flow.
- Ensure project assumptions are reviewed *before* project commences.
- Operating income impact is the acid test:
  - Validate assumptions.
  - Determine what investments are associated with the project (should be low or no capital).
- Update project information to reflect what is known.

- Update financial expectations upon project completion.
- Validate revenue, operating income, or cash flow claims.
- Include in unit cost updates.
- Include in next rolling estimate.
- Periodically review control plans.
- Variance analysis.
- Schedule checks of previously completed projects (self-audit).
- Linkage of financials to performance measures.
- Project-level detail is crucial to success.
- Relate performance measures to P&L and balance sheet items.
- Offline analysis may be needed to understand COPQ levels and impact to Financial Statements.
- Achieving targets depends upon building a path to plan.
- Update information using actual/estimate data.
- Periodic review of financial methodology is required for consistent project valuations.

As you can see, the role of finance is not trivial in deploying Six Sigma. In fact, at the railroad company, CSX, the CFO was the Initiative Champion. Finance can be involved in all phases of selecting and completing the Six Sigma projects. These are a recap of other opportunities for financial involvement:

- **Project Identification**
  - Strong participation in opportunity assessments.
  - Strong participation in project selection.
- **Initiating Projects**
  - Determine incremental investment.
  - Validate assumptions in project charter.
- **Monitoring Projects**
  - Participate in periodic reviews with project team leaders.
  - Participate in periodic reviews with business leaders.
- **Completing Projects**
  - Determine results versus project plan.
  - Update forecasts/estimates.

The finance resources must be willing to ask this question every time a Six Sigma Team reports a financial savings: "Whose budget do I reduce by this much next year?" In the hands of excellent financial resources, Six Sigma becomes a valid bottom-line initiative with lasting impact on the company's performance. But you can see that there is significant predeployment activity required to launch Six Sigma on solid financial ground.

SIX SIGMA HUMAN RESOURCE SUPPORT

Because of the heavy focus on training and professional development, the HR function is expected to be heavily involved with the Six Sigma deployment. To add to the development focus, Six Sigma also adds new positions—Master Black Belts, Black Belts, Green Belts, Initiative Champions, Deployment Champions, and Project Champions. Creating job descriptions and career paths for these new roles is crucial to the success of the deployment. The following lists address roles and responsibilities and training requirements for the Human Resources function.

- **Roles and Responsibilities**
  - Assist in personnel selection.
  - Develop workforce practices that enable Black Belts and Green Belts to achieve high performance in their workplace.
  - Work with Executives and Managers to establish appropriate recognition and rewards.
  - Develop Six Sigma education and training support that contribute to employee performance.
  - Assist in establishing appropriate career plans and career paths for MBBs, Black Belts, and Green Belts.
  - Assist MBBs, Black Belts, and Green Belts in developing their training plans.
- **Training Requirements**
  - Attend Six Sigma Business Team and Deployment Champion training.
  - Attend a one-day HR Workshop.
    - Improvement program organization structure.
    - Recognition and rewards.
    - Appraisal and performance.
    - Retention and career-path planning.
    - Change management considerations.
    - Communication plan.
    - Team effectiveness.

The roles of the HR function are wide and comprehensive. It's almost like starting a new business within the company that is expected to develop people and generate revenue. The HR function will be effective when the HR folks learn the internal workings of Six Sigma. In fact, some of the best Green Belt projects come out of HR.

SIX SIGMA PROJECT TRACKING

When you have hundreds of project teams in action across your company, keeping track of that activity and, more importantly, the results becomes very difficult very quickly. You will be anxious to report the results to Wall Street, and the numbers you present had better be accurate. The backbone of the Six Sigma infrastructure is an automated project tracking system.

A system such as this will allow you to assess the status of your initiative in close to real time. You will know how many projects are active, how many are completed, and how many are stalled, and forecast the business impact of the ongoing projects. All this is necessary information if you plan to report results to Wall Street. A project tracking system can accomplish the following:

- Tracks company-wide program.
- Tracks goals versus actual performance.
- Highlights accountability.
- Decreases reporting time.
- Improves communications.

You will know this for all businesses and functions, and Champion by Champion. Chapter 14, "Defining the Software Infrastructure: Tracking the Program and Projects," is focused on a project tracking system. The good news is that the system will be available for any change initiative that you drive in the future.

SIX SIGMA STEERING TEAMS

We've already seen the discussion of the Executive Steering team and the steering team of Deployment Champions with the Initiation Champion at the head. But there will probably be other steering teams needed to keep Six Sigma on target. These steering teams are focused on specific and relatively narrow applications of Six Sigma and Lean. These steering teams would be deployed as needed. The steering team takes on accountability for the new applications and provides leadership and direction for their part of the program.

While at AlliedSignal, the chemical laboratories at the chemical plant were interested in a laboratory version of Six Sigma. I spoke with the Laboratory Council and we agreed that I would launch a pilot version of the program and the council would be accountable for the teaching and deployment. That's exactly what happened, and we realized a huge improvement in lab performance. Other steering team examples are as follows:

- Business Process Improvement Steering Team
- Product Develop and R&D Steering Team
- Sales and Marketing Six Sigma Team
- Manufacturing Steering Team
- Supply Chain Steering Team
- Human Resources Steering Team
- Equipment Reliability Steering Team
- SAP Implementation Steering Team

Steering teams pick up the responsibility for driving a training plan for their constituents and the accountability for their programs. The teams make life much easier for the Six Sigma Initiative Champion because he or she now has the businesses and functions taking ownership of the program.

He or she may also use the steering teams as a set of board of directors to help the Initiative Champion lead the program strategically. Figure 8.2 depicts the organization of the Six Sigma Steering Teams, all of which report to the Six Sigma Steering Team lead by the Initiative Champion.

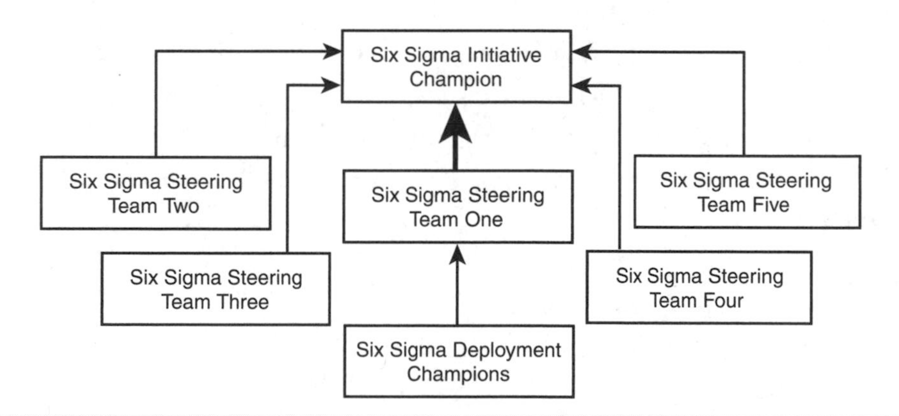

Figure 8.2 The Initiative Guiding Coalition. The Six Sigma Initiative Champion established various Steering Teams, with the first team consisting of the Six Sigma Deployment Champions from each division.

SIX SIGMA CERTIFICATIONS

Because of the depth and length of Six Sigma training for all the Belts and the accountability for results, certification is usually a requirement. Each company defines it own certification. The Six Sigma Steering Team will guide the development of certification requirements to ensure consistency across the corporation.

At this time, there is not a national standard for certification and there shouldn't be. Six Sigma is a business program, and it should be up to the businesses to determine success as a Black Belt or other Belt. The following lists summarize the usual certification requirements for Master Black Belts, Black Belts, and Green Belts.

- **Any Six Sigma "Belt" is certified on two criteria:**
  - Criterion one: Business results.
  - Criterion two: Mastery of tools.
  - Sign-offs:
    - Business results: Business or functional leader and finance director.
    - Tool mastery: Champion and/or Master BB.
  - Elements:
    - Check-off sheet.
  - Final report.
  - Final formal presentation at a high level:
    - Implemented control plan owned by the business or functional leader.
    - Reward and recognition.
- **Master Black Belt candidates are chosen based on**
  - Demonstrated ability in achieving results.
  - Interest in driving the Six Sigma program deployment.
  - MBB development program is tailored to needs of candidate and company, and generally contains
    - Assignment of a mentor.
    - Additional (off-site) training.
    - Statistical methods.
    - Leadership.
    - Setting of specific goals.
  - 12- to 24-month process, depending on capability and prior experience of candidate.

The following is an example of the final certification report check list for a Black Belt or Green Belt candidate. This example represents certification requirements for the manufacturing operations Belts.

Black Belt or Green Belt Certification Checklist

These two sections provide guidelines for Six Sigma Black Belts or Green Belts to be certified. The first section details the items to be delivered to the Director, Six Sigma. The second section provides Six Sigma Champions and Master Black Belts a checklist for Six Sigma tools.

Certification Deliverables

- Project Cover Sheet
- Signatures of Plant Manager, Project Champion, and Master Black Belt
- Copy of Final Report
- Copy of PowerPoint Presentation

Six Sigma Tools Checklist

1. Measurement
    - Process Map (Required)
    - Cause and Effects Matrix
    - Measurement Systems Analysis (Required)
    - Capability Study (Required)
2. Analysis
    - Multi-Vari Study
    - Correlation
    - Failure Modes and Effects Analysis (Required)
3. Improvement
    - Evolutionary Operations
    - Plant Experimentation
    - Fractional Factorial
    - 2K Factorial (Required if Appropriate to the Project)
    - General Factorial
    - Response Surface Methodology

4. Control
   - SPC
   - Control Plan (Required)

SIX SIGMA INFRASTRUCTURE SUMMARY

The Six Sigma infrastructure provides the necessary flow of information from the Executive Team to the businesses. Figure 8.3 shows the infrastructure in action. The Initiative Champion leads the Six Sigma Steering Team, made up of at least the Deployment Champions from each business. The Deployment Champions each have their own Steering Team for the businesses made up of the Project Champions and Master Black Belts. The Project Champions lead the Black Belts and Green Belts in their areas.

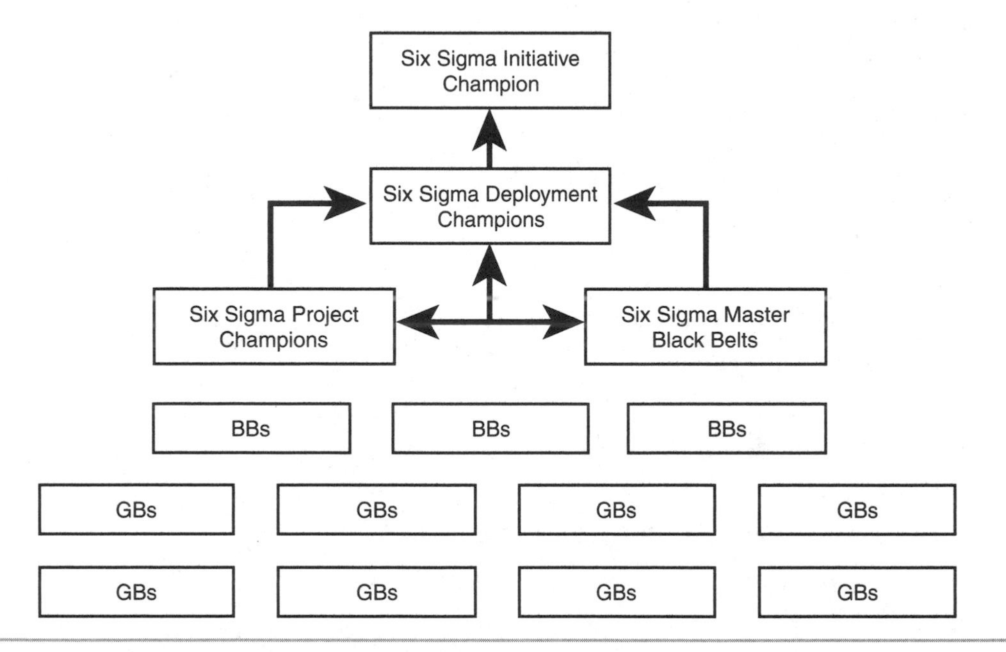

Figure 8.3 The Six Sigma Guiding Coalition. The Six Sigma Initiative Champions lead the Deployment Champions. The Deployment Champions lead the Project Champions. The Project Champions lead the Black Belts and Green Belts.

Figure 8.4 shows the project executive triad. For every Six Sigma project, the Project Champion, Master Black Belt, and Black or Green Belt work closely together to identify, charter, resource, and manage the results of each projects.

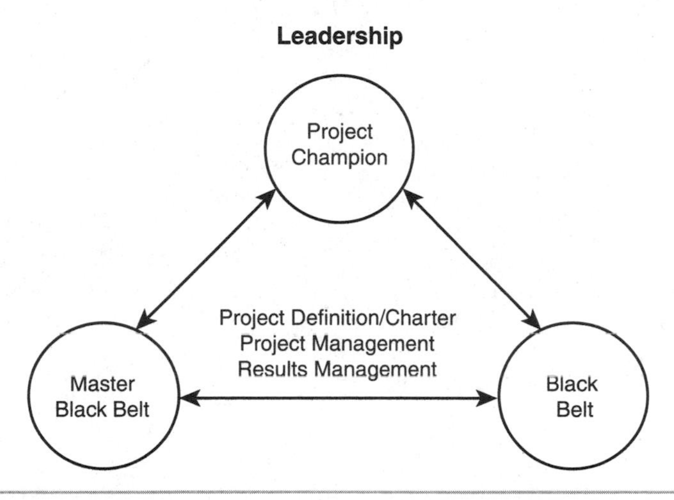

Figure 8.4 The Six Sigma Project Guiding Coalition. The Project Champion, Master Black Belt, and Black Belt work closely on managing each Six Sigma project for results.

The infrastructure becomes a tight network focused on results. Everyone is clear about their roles and responsibilities and their accountabilities. This is the way all change initiatives are successful.

9 Committing to Project Selection, Prioritization, and Chartering

For any initiative, the greatest challenge is identifying what to work on first. Business success is largely a function of focus and prioritization. From Labovitz and Rosansky's excellent book, *The Power of Alignment*, the challenge is, "The main thing is to keep the main thing the main thing!" To make the main thing the main thing comes to play in driving the philosophy of project selection. The dominant strength of any Six Sigma deployment is that it allows you the ability to identify great projects upon which to work.

The old "garbage in, garbage out" adage certainly applies here. Many TQM deployments fell short for the lack of assigning TQM teams something important to work on. Many companies left it up to the teams to figure what projects they would address. The teams didn't have the big picture and tended to work on what was important to them instead of what was important to the business.

Motorola did a great job in helping organizations prioritize projects. By insisting that everyone track defect rates (metric: defects per unit) on a minute-to-minute basis, every organization had a prioritized list (Pareto) of the most frequently occurring defects. The trend chart portraying a positive or negative trend in defects per unit was carefully reviewed at several levels of leadership. The project selection was fairly easy. Priority projects were those that addressed the most frequently occurring defects or the defects that were most important to customers.

As a production manager in Motorola, and a statistician with a Ph.D., I felt I was not really ready to run a large production line (three-shift operation with over 250 operators). I took the job because, in my heart, I felt I had to show the efficacy of running a manufacturing operation using Six Sigma. Although fellow statisticians thought I was crazy, those two years turned out to be the most interesting and successful of my career. Here's what I did.

Based on the production line's Paretos of defects, I assigned my manufacturing engineers and technicians to specific process areas. I had each engineer and technician list the top defects in their process areas, along with potential project ideas to eliminate those defects. We reviewed the proposed projects as a group and drew our line in the sand. We then estimated the expected percentage reduction of each defect. I forecasted what our total defects per unit would be by the end of the year. In a weekly quality meeting, the team (including supervisors) would review the status of each Six Sigma project.

Now, I was nervous. I had never done this before and neither had any other production manager in the plant. Would this really work? Had the right projects been selected? Could I manage this improvement process? Well, the end of the story is positive. We ended up beating the forecast by a significant margin, and we looked pretty good.

Luckily we had a great defect tracking system, so we could determine within about 20 minutes if an experimental process change was working or not. We were in the position to analyze the outcome of planned experiments very quickly. But, in the end, giving my group of engineers and the line operators (they were assigned to high-performance work teams), a sane way to prioritize and go after big issues was the ticket. To create a system centered around one or more metrics will be your challenge.

Your challenge is actually greater than mine was. You've got to commit to creating a project selection system that covers almost every aspect of your business. Out of that system will emerge a new list of great projects as the new projects are completed. Along with that system, you will also have to develop an IT system that allows you to track the progress of ongoing projects across your company. Luckily, there are some excellent systems already available, as discussed in Chapter 14, "Defining the Software Infrastructure: Tracking the Program and Projects."

Six Sigma Leadership Roadmap

My company has developed a five-step Six Sigma leadership roadmap, as presented in Figure 9.1. The inputs are the strategic and annual operational plan goals and target, and the output is making the numbers at the end of the year. The five steps in the roadmap are as follows:

1. Select the right projects.
2. Select and train the right people.
3. Plan and implement the Six Sigma projects.

4. Manage Six Sigma for excellence.
5. Sustain the gains.

SELECTING THE RIGHT PROJECTS

It's no coincidence that the first step addresses project selection. The final results of each year's Six Sigma efforts are directly related to the quality of the projects inputted into the leadership roadmap. This roadmap will be discussed in detail in Chapter 15, "Leading Six Sigma for the Long Term," and each of the preceding steps has its own roadmap. The steps in the roadmap for the first step, *selecting the right projects*, are as follows:

1. Clarify the big picture.
2. Establish the productivity baseline.
3. Prioritize projects based on customer value, resources required, and necessary timing.
4. Select key projects with leadership buy-in.
5. Check accountability:
   a. Business level.
   b. Personal level.

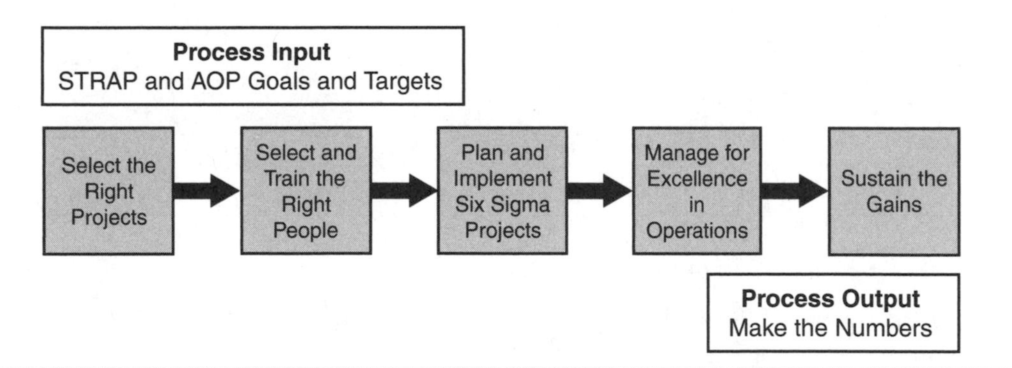

Figure 9.1 The leading Six Sigma leadership roadmap—outputs and inputs.

With this roadmap in mind, let's review the importance in the business model of linking internal activities with the external realities. Linking the internal Six Sigma projects with the company's strategic plan ideally completes this linkage. We'll review a project example list from Chapter 2, "The True Nature of Six Sigma: The Business Model," because it relates directly to project selection. Table 9.1, with the nine categories of target operational metrics, includes a list of example project categories. We'll call these operational metrics *Business Critical Ys* at this point. Because the operational metrics are outputs from your system, selecting the right outputs provides you with a list of Critical Ys.

As you can see, these project categories cover a wide number of functions within the organization, and all cover a pretty wide scope. For example, the project category titled *improve product flow* can easily consist of many projects focusing on numerous process steps or multiple product lines.

Project Clusters. I will refer to these project categories as *project clusters* correlated to a set of projects. Each project cluster will have several projects that are identified and prioritized. The challenge is to prioritize these nine operational metrics (Critical Ys) into, say, three or four, then prioritize each of the project clusters within each operational metric, and then prioritize the projects within each of the top priority clusters. The operational metrics (Critical Ys) with associated project clusters are as follows:

1. **Gross Sales**
    1. Reduce sales price errors.
    2. Improve capacity productivity.
    3. Optimize mix of products sold.
    4. Increase price through good product design.
2. **Discounts**
    1. Reduce special sales allowances.
    2. Reduce contract discounts.
3. **Cost of Sales (Material, Labor, Overhead)**
    1. Improve product flow.
    2. Lower supplier costs.
    3. Improve quality in production.
    4. Improve quality in suppliers.
    5. Improve payables terms.
    6. Reduce inventory owned.
    7. Redeploy assets.
    8. Improve equipment uptime.
4. **Warranty**

1. Improve manufacturing processes.
2. Improve source of components.
3. Reduce testing.
4. Streamline warranty process.
5. Improve warranty forecasting.
6. Eliminate abuses of warranty system.
7. Improve design process to improve process and product quality.
8. Fix issues quickly to satisfy customers.

5. **Engineering Expenses**
   1. Reduce time to develop a product or service.
   2. Reduce required product testing.
   3. Improve cycle time of engineering change notices without impacting quality.
   4. Better utilize partners and suppliers for better designs.
   5. Improve technology development.
6. **Marketing Expenses**
   1. Reduce discounts.
   2. Eliminate low or no margin businesses.
   3. Improve capacity productivity.
   4. Improve processes for communicating product or service volumes.
   5. Improve forecasting process.
   6. Improve voice of the customer data collection.
7. **Administrative Expenses**
   1. Improve administrative flows.
   2. Reduce defects in administrative processes.
   3. Reduce depreciation, taxes, and insurance.
   4. Eliminate central inventory holdings.
   5. Reduce the number of suppliers.
8. **Interest**
   1. Improve management of cash.
   2. Reduce inventory levels.
   3. Improve receivable process.

4. Improve payable process.
5. Minimize fixed-asset investments.

9. **Taxes**
   1. Optimize tax process to obtain tax advantages.
   2. Prevent overpayment of taxes.
   3. Optimize inventory levels.

An example of Critical Ys and project clusters are lifted from Chapter 2. Let's say the Critical Ys for a company have been determined to be operating margins, cash flow, return on capital, revenue, and return on investment. Going back to the Y = Function (Xs) formula, our Ys will be the Critical Ys and the Xs will function as project clusters:

- $Y_{\text{(operating margins)}} = f$ (revenue, product costs, business costs)
- $Y_{\text{(cash flow)}} = f$ (profits, working capital)
- $Y_{\text{(return on capital)}} = f$ (volume, average selling price, discounts)
- $Y_{\text{(revenue)}} = f$ (volume, average selling price, discounts)
- $Y_{\text{(return on investment)}} = f$ (profit, investments, net assets, asset turnover)

So the project clusters for the Critical Y of operating margins will be revenue, product costs, and business cost. The company will now look for potential projects that address those three areas. Projects addressing product costs may include projects in yield improvement, reduction of scrap, and improvements reducing cost-of-poor-quality. So, at the end of the year, your company should be able to link projects to project clusters and to the Critical Ys, respectively. A summary statement for improving operating margins, at the beginning of the year, follows.

Improvments in operating margin will include bottom-line impact of capacity expansion (growth) or cost reduction (productivity) versus prior year. So, a bunch of projects addressing the two project clusters (capacity expansion and productivity) should be identified and chartered early in the year. Other examples of Critical Ys linked to the strategic goals of improving growth, productivity, and working capital are the following:

- **Growth**
  - Revenue due to mergers and acquisitions.
  - Revenue due to new products.
  - Sales volume.

- **Productivity**
  - Capacity/productivity.
  - Cost of poor quality.
  - Rolled throughput yield.
- **Working capital**
  - Inventory turns.
  - Days sales outstanding.
  - Finished goods inventory.

Figure 9.2 illustrates the linkage among strategic initiatives, Business Critical Ys, project clusters, and projects. Using this methodology allows each project to be linearly mapped directly to a specific strategic initiative. Figure 9.2 follows one line from the strategic initiative of working capital to the Critical Y, projects cluster, and a set of projects for one project cluster.

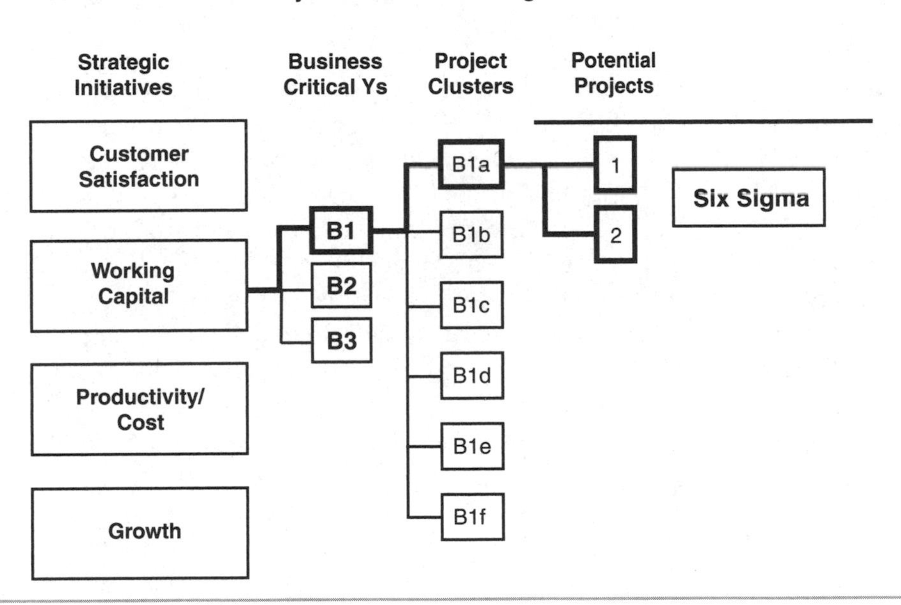

Figure 9.2 Strategic initiatives linked to Business Critical Ys, project clusters, and projects. Each of the four strategic initiatives would have a similar tree alignment.

Taking the highlighted line in Figure 9.2, Figure 9.3 shows what one project line would look like for the strategic initiative of working capital. This figure shows that working capital might have a Critical Y of reducing working capital to $10.7 million, a project cluster addressing the customer order process, and a project with the goal to minimize order adjustments. Each project cluster would have a set of identified potential projects tied to it.

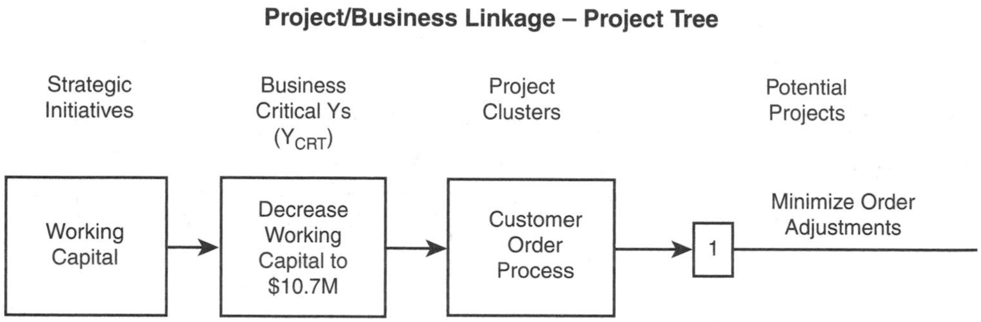

Project/Business Linkage – Project Tree

Strategic Initiatives	Business Critical Ys (Y_{CRT})	Project Clusters	Potential Projects
Working Capital →	Decrease Working Capital to $10.7M →	Customer Order Process →	1 Minimize Order Adjustments

Figure 9.3 Strategic initiatives linked to Business Critical Ys, project clusters, and projects. Each project has a line of sight to a strategic initiative.

Figure 9.4 shows a more detailed example. Suppose we look again at this company's strategic initiative of working capital level. They have defined three Business Critical Ys (sometimes known as *goals*): net working capital turns, cycle time, and decreased working capital. The business teams have identified some projects clusters. The next step would have the different functions define projects that address the project clusters. All those projects would directly impact cash and working capital levels. There could easily be 40 or more projects addressing this proposed package.

So, now we review the *project selection and prioritization roadmap* that includes some steps from all five steps of the project selection roadmap. This is a high-level overview of the actions that must be accomplished several times a year to ensure that all Six Sigma projects are the right projects for the company. Each of these steps will be covered in detail with examples.

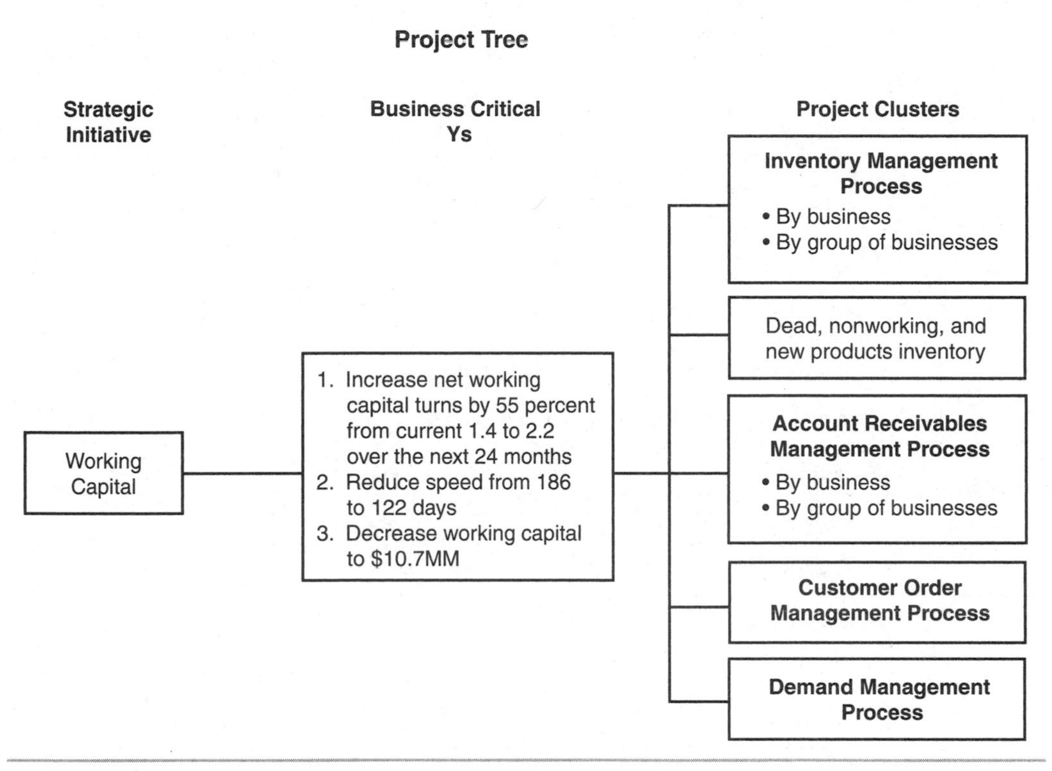

Figure 9.4 The linkage between the strategic initiative, Business Critical Ys, and project clusters. The next step is to identify projects for each project cluster.

PROJECT PRIORITIZATION AND SELECTION ROADMAP

You will be using the roadmap shown in Table 9.1 to develop your project selection system that will be used throughout your company. There might even be a standard operating procedure to describe project selection. The final project selection system will consist of steps in this chapter and steps that have worked in your company before. Because project selection is so important to Six Sigma, or any initiative, creating a process and a system will result in consistency across your organization and improve the probability of each function successfully identifying projects that will line up with your strategy and your external realities.

Table 9.1 Project Prioritization and Selection Roadmap

1.	Establish long-term strategic plan.
2.	Establish and review business priorities (Critical Ys) for the year.
3.	Brainstorm clusters for each business priority (Critical Y).
4.	Prioritize the project clusters.
5.	Determine initial project priorities within each cluster using a decision (cause and effects) matrix—draw the line.
6.	Perform an initial project screening using predefined project filters—draw the line again.
7.	For top projects, complete short-form of project charters:
	a. Process.
	b. Problem.
	c. Metrics—baseline, entitlement, goal.
	d. Business impact ($).
	e. Project sponsor, team leader, team members, cost for expenses, capital, and resources.
8.	Review project charter short-forms and make final selections.
9.	Develop final charter with team leader (Black Belt, Green Belt).
10.	Decide whether project is Six Sigma or other.
11.	Review final charters for management approval.

Because project selection is so important during the first 90 days of your Six Sigma launch, project selection makes up a significant amount of launch activity and time. Table 9.2 presents the short-form of the timetable provided in Chapter 7, "Defining the Six Sigma Project Scope." This timetable for the first 90 days highlights the milestones related to project selection. Of the 11 milestones, the five boldfaced milestones are directly related to project selection. Although good project selection relates to all 11 milestones, over 50 percent of the milestones are focused on project selection.

Table 9.2 Key Actions for the First 90 Days of Your Six Sigma Launch*

1.	Identify time line for deploying:
	• Black Belts
	1. Operations, Services, Product Development.
	• Green Belts
	1. Ops, Services, Product Development.
2.	**Determine project filters at the plant and business level.**
3.	**Identify potential projects.**
4.	Identify Sponsor and Black Belt candidates.
5.	**Identify the priorities for the projects and assign to the appropriate Black Belts(s).**
6.	**Complete a charter for each project, including financial impact.**
7.	Determine the review methods and frequency for your plant and business.
8.	Determine if the resources and leadership commitment are in place to make the project(s) successful.
9.	Determine the communications methods for your plant and business.
10.	**Implement IT Project Tracker.**
11.	Develop reward and recognition guidelines.

* Bold-faced milestones are directly related to project selection.

We will now go through the roadmap for the leadership step, select the right projects. The first step is: *Clarify the big picture.*

Clarify the Big Picture. Clarifying the big picture is the process of establishing linkages. Going to the business model, we must link the issues associated with our external reality to clearly stated financial target and coordinated internal activities.

The two actions for this step are (1) to identify the business strategic initiatives and (2) to translate these into primary operational objectives for all levels and divisions of your organization. Along with these, financial and performance targets (Critical Ys) will be disseminated. So if productivity improvement, cash flow improvement, and cost improvement are the important things (Critical Ys), these must be made clear to the company.

Any projects that don't address these three strategic areas are immediately called into question. Clarifying the big picture can be simple. Paul Norris, CEO of WR Grace, simply put one slide on the overhead projector: $56 Million. That's the level of pretax profit he was looking for at the end of the year from Six Sigma actions. Everyone in the room was aligned with Paul. The level of urgency noticeably increased. Figure 15.5 shows a summary diagram of Step 1 of the leadership.

Let's suppose we have a manufacturing company whose main strategy centers around establishing operational excellence. Its primary strategic goals are as follows:

- Improve process reliability.
- Increase capacity.
- Reduce operating costs.
- Transform the culture.

Figure 9.5 illustrates going from strategies to Business Critical Ys to a list of projects for each Critical Y. I have omitted the set of project clusters for each Critical Y. You'll notice that all projects are not necessarily Six Sigma projects. For example, the project, training Black Belts and Green Belts, is more of an action than a process improvement project. This list of projects, if successful, will move this company forward in its challenge to attain the strategy.

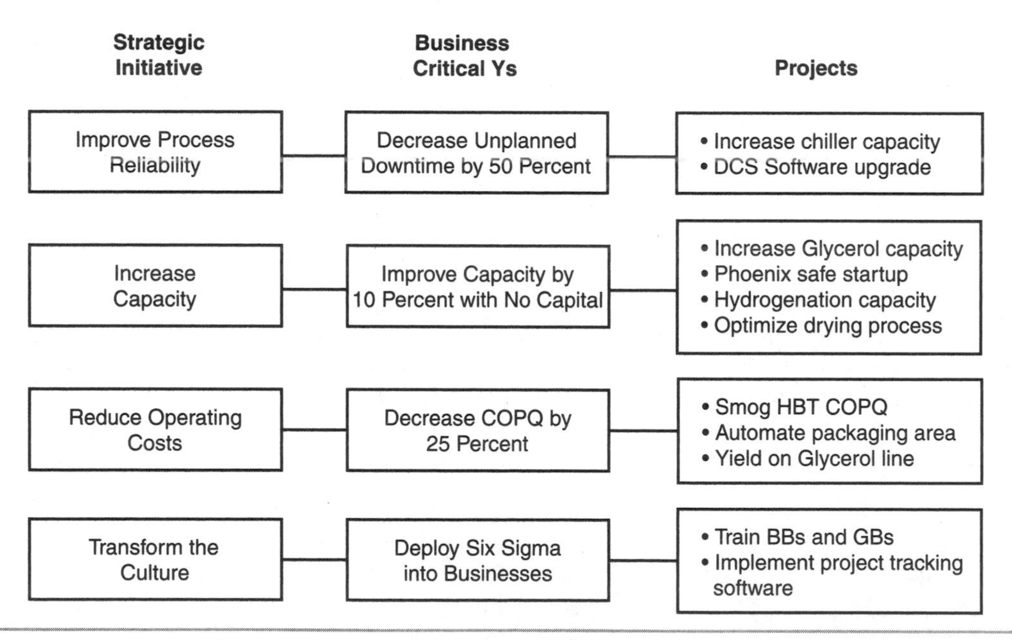

Figure 9.5 The linkage between the strategic initiative, business Critical Ys, and projects. The projects are aligned with project clusters for each Critical Y.

TYPES OF PROJECTS: TOP-DOWN AND BOTTOM-UP

Six Sigma projects usually fall into two primary categories: Top-Down and Bottom-Up. The Top-Down projects are projects that are discovered by starting with the strategic initiatives and the annual operating plan. These are the projects that will move the company forward strategically. Bottom-Up projects are focused tactically on problems that need to be solved immediately. Bottom-Up projects address the areas in which the company is feeling pain. For example, for a sold-out product (you can sell everything you make) not meeting customer demand, any increase in manufacturing capacity will be immediately welcome.

Top-Down Projects. In the ideal world, all Six Sigma projects would be top-down. All projects would be driven by the strategic plan and the annual operating goals around that plan. Every successfully completed project would place the company that much closer to executing the strategic plan. Because there is a history of CEOs getting fired largely due to their inability to execute strategy, this core capability is crucial to the success of the company and its CEO and senior leaders.

To select Top-Down projects is to identify the relevant business processes that, if improved, will move the company forward strategically. To initiate the Top-Down project selection process, two steps are completed for each of the selected business processes: (1) Calculate the baseline process performance; and (2) Estimate the performance entitlements.

The *process baseline* represents the current process performance on a set of the Critical Y's and the *process entitlement* is the theoretical performance on the same metrics if everything ran perfectly. Six Sigma projects reflect the actions necessary to close the gap between baseline and entitlement. For example, the day's sales outstanding (DSO) might currently be 52.2 days.

Entitlement is defined as the best conceivable performance of a process on any number of variables. For example, the entitlement of yield is 100 percent, of cost of poor quality is zero, and of capacity might be the nameplate or scientifically theoretical performance of a process according to the equipment manufacturer.

By moving from the baseline of 52.2 days to the entitlement of 25 days, we would put about an additional $32 million to the bottom line every year. That would convince us that reducing DSO from 52.2 days to a target of 38.6 days (closing the gap between baseline and entitlement by 50 percent) is a great Six Sigma project. Figure 9.6 gives examples of entitlement estimates for a manufacturing process. We are moving toward perfection, and entitlements help us understand the business worth of that effort. Figure 9.7 shows how baseline and entitlement can yield process performance goals.

Manufacturing Process Baselines and Entitlements

Metric	BSL	NOW	Goal	Entitlement	Units
Defects/ Unit	.05	.035	.025	.003	DPU
Scrap	520K	320K	120K	60K	$/A
Downtime	20%	18%	10%	3%	$/A
Rate	1.4K	1.6K	2.2K	3.0K	Units/Hr

Figure 9.6 An example of manufacturing process baseline (BSL) performances, process entitlements, and setting goals to close the gap between baseline and entitlement.

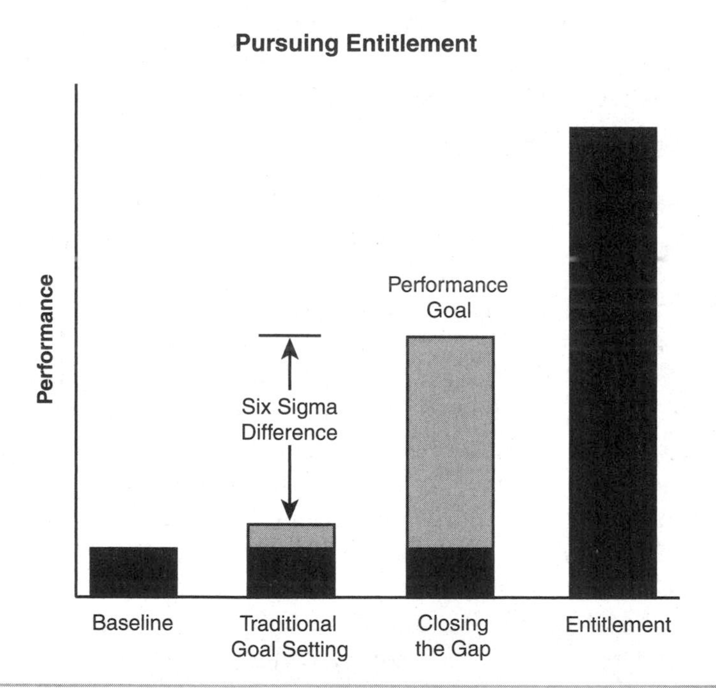

Figure 9.7 Comparing baseline performance (left column) to process entitlement (right column) and setting goals to close the gap between baseline and entitlement by 50 percent.

Figure 9.7 shows that, by knowing process baseline performance and estimating entitlement, goals can be aggressively set. A good rule is to close the gap between baseline and entitlement by 50 percent, which is a tough but doable goal.

Bottom-Up Projects. Significant known problems and opportunities can often be identified immediately. Every business confronts chronic problems with their business processes everyday. I was in a rehabilitation hospital and I repeatedly heard, "If we could just fix the scheduling system, my job as a therapist would be much easier."

Defects in the scheduling system either had patients showing up at the wrong place and the wrong time, or therapists were missing. These mistakes had a huge impact on the operation of this hospital. Here are some other examples of chronic problems:

- Excessive yield loss or product downgrades.
- Customer dissatisfaction from chronic late shipments.
- Inability to penetrate new market segments due to poor process capability.
- Order backlog due to suboptimal capacity utilization.

Each of these issues can be formulated into projects by determining the underlying business/operating process that is performing poorly, and the financial opportunity that would result from improved process capability. To get a good start on identifying the ubiquitous Bottom-Up projects, establishing the baseline process performance will help identify those black holes of process performance. Suppose we have a process that has three steps, and we baseline first-pass yield, cost of poor quality (COPQ), and capacity.

Perfect first pass yield is 100 percent. Anything less than 100 percent means we're generating scrap or time penalties. COPQ represents the costs of any work in the manufacturing process that is not directly related to adding value to the product. Scrap, rework, inventory, defects, and testing are all examples of COPQ. Every dollar of COPQ we save goes directly to the bottom line. And, finally, the process step that has the least capacity in units per day is the bottleneck. Improvements in capacity will raise the productivity of the entire line.

We can see from Figure 9.8 that process step 1 has the worst yield (DPU = .35), process step 2 has the worst COPQ ($0.30 per unit) and, finally, process step 3 has the worst capacity (500 units per day). These are three potential projects. The next step is to set goals and quantify the financial impact of hitting those goals. Figure 9.9 shows each of the three projects analyzed financially. We see that the three projects have a potential to bring about $2 million to the bottom line. Not a bad set of projects.

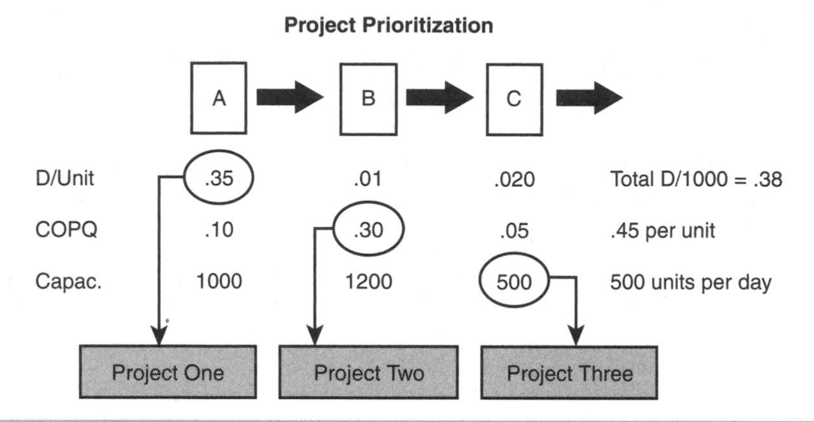

Figure 9.8 Using the baselines for defects per unit, cost of poor quality, and capacity to identify Bottom-Up projects.

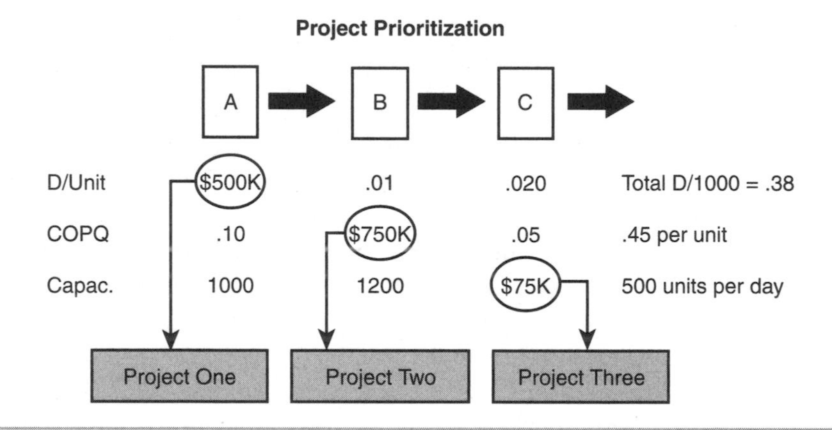

Figure 9.9 Assessing the projects identified from baseline performance for financial impacts.

Project Selection Trends. Initially in a Six Sigma deployment, most of the projects selected tend to be Bottom-Up projects because we know our current situation better than our future situation. But as the Six Sigma deployment progresses, Figure 9.10 shows the evolution of Top-Down and Bottom-Up projects. You can see that the percentage of projects that are Top-Down tends to increase year over year. This is when the real bang for the Six Sigma buck comes.

Your company will become more forward looking because it is operating smoothly day-to-day due to many of the Bottom-Up projects that have been completed.

AlliedSignal, in 1995, brought over $350 million to the bottom line largely from Bottom-Up projects. Design for Six Sigma was initiated in early 1996, which resulted in a large number of the Six Sigma projects being Top-Down.

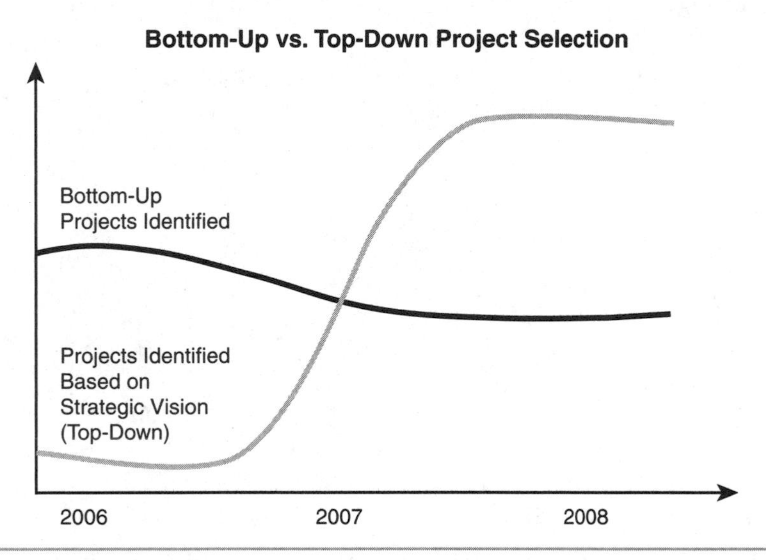

Figure 9.10 As the Six Sigma initiative matures, the proportion of Top-Down strategic projects increases.

PRIORITIZING PROJECTS

Here's the situation. A specialty chemical company (I'll call Chemex) is having an improvement planning meeting to identify improvement projects for next year. This $100 million business identified over 125 projects. Everyone was a little nervous. Most thought, including me, "We don't have the resources or time to do 125 projects."

Chemex made the decision to allow my consulting group to meet with the team and go through a prioritization exercise. The effort worked great. They went from a shotgun approach of 125 projects to a precise set of 25 projects. Every project was linked directly toward their strategic plan, and they determined ahead of time that they had the resources to commit to this set of projects.

The toughest part in launching your first wave of Six Sigma Black Belts is sending each student to training with an important project that is doable in less than six months. Each project also has the necessary resources (people and capital) and is tracked by the functional leadership. Any organization can brainstorm 10 flip charts worth of potential projects. But the prioritization is the tough part.

If you use the rule of 80/20 (only 20 percent of the project ideas are really important), developing and committing to a system of prioritization is essential. There's a big difference in committing to complete 25 projects when compared to 125 projects. Giving your organization the feeling that their time and energy are being dedicated toward accomplishing something important is what excellent leadership is about.

Moving from Strategy to Projects. Chemex, our fictitious chemical company, has decided that its primary strategic initiatives are the following:

- Customer satisfaction (On-time delivery)
- Cash/working capital
- Productivity/cost
- Revenue growth

The leadership team then defined the Business Critical Ys with some potential project clusters for each of the four strategic initiatives. Figure 9.4 shows an example of the Critical Ys and project clusters for the cash/working capital strategy.

Project Prioritization Matrices—The Top-Down Matrix. Each business division or function would brainstorm projects for each project cluster or strategic initiative. The Chemex organization created a preliminary list of 125 projects. Each project group (division or function) then evaluated the projects via a decision matrix similar to a QFD matrix. Figure 9.11 shows the working matrix.

Place Projects in the Prioritization Matrix

Rating of Importance →	5	7	7	10	
Project Ideas	Customer Satisfaction	Cash/ Working Capital	Productivity/ Cost	Growth	Total
2095 System Uptime	3	1	3	1	53
Aqueous Scrap Reduction					0
Former Rate Improvement					0
Gross Spend Targets					0
HFE Yield Improvement					0
HQ-115 Yield Improvement					0
HX-686 Scrap Redux					0
M2000 Productivity Improve					0
New Product Intro Process					0
S&OP Process Definition					0

Project Opportunities → (project ideas rows); Critical Ys → (strategic initiative columns)

Figure 9.11 Top-Down prioritization matrix scoring potential projects on how well they impact the strategic initiatives. The total score for this example is the sum of the products of importance rating and project impact.

The four strategic initiatives are displayed horizontally across the top of the matrix. The potential project opportunities are displayed down the left column. The purpose of this matrix is first to prioritize (or rank) the potential projects by their total potential impact of the four strategic initiatives. Some projects will impact more than one initiative. For example, an increase in manufacturing capacity can impact growth, customer satisfaction, cash, and productivity.

The first step is to rank each strategic initiative on a 1–10 scale, with 1 being not very important and 10 being extremely important. Figure 9.11 indicates that Chemex looks at Growth as the most important initiative (Score = 10) and Customer Satisfaction as the least important (Score = 5).

The next step is to rate each project's impact on the four strategic initiatives. A simple scoring system will yield numbers of 0, 1, 3, or 9. The scoring criteria are

- 0 = Not at all related to this initiative.
- 1 = Weak impact on this initiative.
- 3 = Moderate impact on this initiative.
- 9 = Strong impact on this initiative.

Then, each project score is multiplied by the score for the initiative and summed over those products. Figure 9.11 shows one project as it's scored and the total. The project, 2095 System Uptime, has a total score of $5 \times 3 + 1 \times 7 + 3 \times 7 + 1 \times 10 = 53$. That way, the projects that impact more than one strategic initiative are favored with higher scores. Figure 9.12 shows the first part of the project list and the scores of each project.

Place Projects in the Prioritization Matrix

Rating of Importance ⟶	5	7	7	10	
Project Ideas	Customer Satisfaction	Cash/ Working Capital	Productivity/ Cost	Growth	Total
S&OP Process Definition	9	3	3	3	117
HQ-115 Yield Improvement	3	1	9	3	115
New Product Intro Process	3	0	1	9	112
M2000 Productivity Improve	3	3	9	1	109
Former Rate Improvement	1	1	9	3	105
Gross Spend Targets	0	3	9	1	94
HFE Yield Improvement	3	0	9	1	88
HX-686 Scrap Redux	3	3	3	1	67
2095 System Uptime	3	1	3	1	53
Aqueous Scrap Reduction	0	3	3	1	52

Figure 9.12 A sample of potential projects rated on total impact of the strategic initiatives. Projects are ranked by total score.

You can see very quickly that the top five potential projects have the highest score because each scores a 9 on at least one initiative, and each scores a 3 on several others. The next step ranks the projects by their total scores, and a line is drawn that separates the best projects from the ones that aren't so good. See Figure 9.13 for the final results. Notice the line delineating the top projects versus other projects that may be considered in the future.

Draw the Line

Rating of Importance →	5	7	7	10	
Project Ideas	Customer Satisfaction	Cash/ Working Capital	Productivity/ Cost	Growth	Total
S&OP Process Definition	9	3	3	3	117
HQ-115 Yield Improvement	3	1	9	3	115
New Product Introduction Process	3	0	1	9	112
M2000 Productivity Improve	3	3	9	1	109
Former Rate Improvement	1	1	9	3	105
Gross Spend Targets	0	3	9	1	94
HFE Yield Improvement	3	0	9	1	88
HX-686 Scrap Reduction	3	3	3	1	67
2095 System Uptime	3	1	3	1	53
Aqueous Scrap Reduction	0	3	3	1	52

Figure 9.13 A line is drawn based on the break in final scores. We have identified the top five projects.

The final step of the Top-Down prioritization technique is to do an initial screening, a short evaluation of the top projects. Figure 9.14 displays an example of an initial screening for the top five projects ordered by the total score, the priority score in this table. This initial screening generates the following information about each project in the form of what we might call "Mini-Charters":

- Project
- Priority Score (Total Score from the Matrix)
- Business Impact—Financial
- Business Impact—Non-financial
- Project Metric
- Personnel Resources—Full-Time Equivalent
- Capital Resources
- Estimated Completion Date

Perform Initial Screening

Project	Priority Score	Business Impact—Financial ($$)	Business Impact—Nonfinancial	Measurement	FTEs	Capital Required	Completion Date
S&OP Process Definition	117	$0.7MM	Lost Sales	OTD	4	None	Jul-01
HQ-115 Yield Improvement	115	$0.8MM	Allows for Business Expansion	% Yield	2.5	$75K	Sep-01
New Product Intro Process	112	$0.3MM	Future Growth	Time to Mkt	5	None	Oct-01
M2000 Productivity Improvement	109	$.26MM	Solvent Free	% Yield	2	$15K	Dec-01
Former Rate Improvement	105	$0.2MM	Return to 5 Day Work Week	lbs/Hr	3.5	None	Apr-01

Figure 9.14 An initial screening of the top five projects based on impact criteria.

You perform this initial analysis to determine if the division or function has the resources to do the project. The analysis also reaffirms the strategic and business importance of each project. The organization can estimate the total financial and strategic impact of each project. Now we are ready to assign each project to a Black Belt, train the Black Belts, and complete the projects.

Project Prioritization Matrices—The Bottom-Up Matrix. That's one way to prioritize—the Top-Down way—starting with strategic initiatives. Next is the Bottom-Up prioritization method. These projects are driven by an immediate tactical need. The prioritization process is the same as the Top-Down method with the different items across the top of the matrix.

Looking at the Chemex example in Figure 9.15, you can see the selection scoring criteria represent a combination of strategic initiative and tactical requirements. For example, Chemex considered improving product throughput and avoiding new capital expenditures by optimizing current processes to be very important to accomplish this year. The project scoring criteria with the criteria scores for the matrix are as follows:

- Operational Dollar Impact (6)
- Process Throughput Improvement (9)
- Time to Complete the Project (7)
- Customer Satisfaction (7)
- Capital Avoidance (10)

So, when we score and rank the projects with the new criteria, we end up with a different set of projects that better meet the immediate business performance gaps. Figure 9.15 shows the list of the selected projects.

	Rating of Importance ⟶	6	9	7	7	10	
	Project Ideas	Dollar Impact (Op Inc)	Throughput Improvement	Time to Complete	Customer Satisfaction	Capital Avoidance	Total
1	2095 System Uptime	9	9	3	3	3	207
2	Aqueous Scrap Reduction	3	3	3	0	9	156
3	Former Rate Improvement	3	9	1	1	9	203
4	Gross Spend Targets	9	3	3	0	1	112
5	HFE Yield Improvement	9	9	9	3	3	249
6	HQ-115 Yield Improvement	3	3	3	3	3	117
7	HX-686 Scrap Reduction	1	3	3	3	9	165
8	M2000 Productivity Improvement	3	3	9	3	9	219
9	New Product Introduction Process	3	1	1	3	1	55
10	S&OP Process Definition	1	3	3	9	1	117

Figure 9.15 Example of project prioritization based on using Bottom-Up criteria.

Project Filters. To obtain a set of reasonably objective scoring criteria, the concept of project filters enters into the prioritization process. Project filters represent a set of project selection rules that almost anyone in the organization can use to evaluate potential projects. Let's look at project filters that Chemex used. These project filters include a list of six scoring criteria along with some number anchors, including the scores 1, 3, and 9.

1. Dollar impact of operating income:
   - a. 9 = > $1MM
   - b. 3 = > $250K
   - c. 1 = < $250K
2. Throughput improvement:
   - a. 9 = > $1MM
   - b. 3 = > $250K
   - c. 1 = < $250K
3. Time to complete:
   - a. 9 = < 6 months
   - b. 3 = 6–9 months
   - c. 1 = > 9 months
4. On-time delivery improvement:
   - a. 9 = >50 percent
   - b. 3 = 25–50 percent
   - c. 1 = <25 percent
5. Capital avoidance:
   - a. 9 = > $5MM
   - b. 3 = $1–5MM
   - c. 1 = < $1MM
6. Other (must specify/define):
   - a. 9 = High Impact
   - b. 3 = Medium Impact
   - c. 1 = Low Impact

The projects are scored using the scoring scheme derived from the project filters and then ranked and the selection line drawn (see Figure 9.16). The list of selected projects from the Bottom-Up prioritization differs from the Top-Down analysis.

As Jack Welch has stated in the past, "You have to eat short term while you grow long term." If a company has serious operating or market issues that are preventing them from performing now, it is best to focus on those issues in their Six Sigma launch. Dynamic leadership supports the organization in understanding what is important now and to act on those priorities.

Draw the Line

	Rating of Importance →	6	9	7	7	10	
	Project Ideas	Dollar Impact (Op Inc)	Throughput Improvement	Time to Complete	Customer Satisfaction	Capital Avoidance	Total
1	HFE Yield Improvement	9	9	9	3	3	249
2	M2000 Productivity Improvement	3	3	9	3	9	219
3	2095 System Uptime	9	9	3	3	3	207
4	Former Rate Improvement	3	9	1	1	9	203
5	HX-686 Scrap Reduction	1	3	3	3	9	165
6	Aqueous Scrap Reduction	3	3	3	0	9	156
7	HQ-115 Yield Improvement	3	3	3	3	3	117
8	S&OP Process Definition	1	3	3	9	1	117
9	Gross Spend Targets	9	3	3	0	1	112
10	New Product Introduction Process	3	1	1	3	1	55

Figure 9.16 Drawing the line to identify the top-rated potential projects.

Here are some other examples of project filters. They're all different because different companies have different priorities.

Project Filter List #1

- *Significant* financial impact to the business:
  - $500K+ Profit Before Tax (PBT) impact within 12 months of project inception.
- *Significantly* improves Rolled Throughput Yield, DPU, COPQ, Downtime, Capacity:
  - 50 percent increase in Rolled Throughput Yield.
  - 50 percent reduction in DPU.
  - 50 percent reduction in COPQ.
  - 50 percent reduction in Downtime.
  - 50 percent increase in Capacity.
- Focuses on *high-volume*/high-risk products.
- Significantly *reduces* field inquiries.
- *Improves* customer service.

Project Filter List #2

- Project significantly *improves* defects/unit, scrap, downtime, capacity:
  - 50 percent increase in First Pass Yield.
  - 50 percent reduction in Scrap.
  - 50 percent reduction in Downtime.
  - 50 percent increase in Capacity.
- Project focuses on *high-volume*/high-risk products.
- Project significantly *reduces* field inquiries.
- Project *improves* customer service.
- Project delivers *$250K* to the bottom line.
- Project can be completed in (*four to six months*).
- Project is focused on a *process* that has *variation*.
- A clear *measurement* (or metric) is defined for the project.

Project Filter List #3

1. Is the project focused on a real problem?
2. Will there be a short-term (six-month) real dollar drop to the bottom line?
3. What is the "probability of success" for the project?

4. Is a Black Belt or Green Belt available to lead the project?
5. Will a majority of the Six Sigma tools be applicable to the project?
6. What resources, financial and human, are required to support the project?
7. Will the project require 12 months or less to complete?
8. Does the project address some part of the "hidden factory?"
9. Will the completion of the project provide process improvement?
10. Was the project identified by the manufacturing staff?
11. Does the project have intangible benefits, customer satisfaction, and so on?
12. What level of customer impact is forecast for this project?

Project Filter List #4

1. Reduce waste by 50 percent.
2. Increase capacity—when > 85 percent sold.
3. Increase service level to > 98 percent on time in full.
4. Reduce customer complaints by 50 percent.
5. Bottom-line impact > $100K.
6. Complete project within six months.

Project Filter List #5

- **General**
  1. Focus on "Quick Hit" projects that will allow for early success.
  2. Projects that lead company to move to "Best in Class" reputation.
- **Customer Growth**
  1. Improve on time performance by 50 percent.
  2. Improve capacity performance by > 20 percent.
- **Technology/R & D**
  1. Dollars/unit time savings of > $100K/yr.
  2. Profitability of > 20 percent.
- **Operations**
  1. RTY/DPU improvement of 75 percent or more.
  2. PPM returns of above 10,000 ppm reduced by 90 percent.

CHECKING AND ESTABLISHING ACCOUNTABILITY: CHARTERING PROJECTS

Project Chartering. The hard part is over. You've now got a list of excellent projects that, if completed, leaves the company performing better than it was before. But the devil's in the details. Your organization has to select Black Belts, train them, and provide the resources necessary for project success.

The step that sets all these actions in motion is chartering the projects. The project charter is a contract among the Six Sigma team, the leader holding the accountability for the project (Project Champion), and the company leadership. By doing a good job with the charter, the important questions are answered before the project is initiated. Expectations are clear, and everyone is locked and loaded. An example of a project charter is displayed in Table 9.3. An example of a Six Sigma project status report is listed in Table 9.4.

Besides the usual boilerplate, there are some sections that are very important. The section titled "objectives" consists of a set of metrics upon which process performance will be based. Section 5—business results in $—is probably the most important section. Understanding the financial potential of the projects takes active participation by the finance function in a company. Each Six Sigma project must have a complete charter.

The Project Champion develops the preliminary charters based on the project list. The Plant, Division, or Function Leadership Team reviews the charters to select the key projects. Key projects have their charter further developed by the Project Champion and the Black Belt. The final charter review is completed with the Champion, Black Belt, and Plant (or functional) Leadership Team. The charter Summary reports are useful in business quarterly reviews.

Here is a list of questions the charter should answer:

1. In measurable terms, what is the project trying to accomplish?
2. Is this project worth doing?
3. What happens if this project fails?
4. Does it fit within the business objectives?
5. Is this a customer-oriented project?
6. How's the scope? Boiling the Ocean or Right Size?
7. What are the specific goals? Stretch targets?
8. Who owns the process? Will they be involved?
9. What's the probability of success?
10. Can we get benchmark information? If so, where?
11. What resources are available to the team?

Table 9.3 Six Sigma Project Charter

Product Impacted		Product Impacted Sales ('99 projection $)	
Breakthrough Leader		Telephone Number	
Process Improvement Leader		Plant/Site	
Start Date		Target Completion Date	

Element	**Description**	**Team Charter**				
1. **Process**	Define the process in which opportunity exists.					
2. **Project Description**	Describe the project's purpose and overall objective.					
3. **Project Scope**	Define the part of the process that will be investigated.					
4. **Objective**	Define the baseline, the theoretical (ideal) target, and the stretch goal for improvement on the primary metrics: Rolled Throughput Yield, Cost of Non-Conformance, and Capacity/Productivity.		Baseline	Goal	Entitlement	Units
		RTY				%
		CONC				$K/yr
		C-P				Units
5. **Business Results: (in 1999 dollars)**	Define the improvement in business performance (e.g., sales and income) that is anticipated and when.					
6. **Team Members**	Define the team members.					
7. **Benefit to External Customers:**	Define the final customer, the benefit they will see, and their most critical requirements.					
8. **Schedule**	Key milestones/dates.	Project Start				
	M—Measurement	"M" Completion				
	A—Analysis	"A" Completion				
	I—Improvement	"I" Completion				
	C—Control	"C" Completion				
	Note: Schedule appropriate Safety Reviews.	Project Completion				
		Safety Reviews				
9. **Support Required**	Define any anticipated needs or any special capabilities, hardware, trials, and so on.					

Table 9.4 Six Sigma Project Status

Product Impacted			Product Impacted Sales ('99 proj $)				
Breakthrough Leader			Telephone Number				
Process Improvement Leader			Plant/Site				
Team Members							
Start Date			Target Completion Date				
Project Specifics	Project Description			Baseline	Now	Goal	Ent
			RTY (%)				
	Objective		COPQ ($K/yr)				
	Projected Business Results	$ Saved to Date	C-P (Units)				

Key Milestones		Date	Status
	1		
	2		
	3		
	4		
	5		
	6		
	7		
	8		
	9		
	10		

Report Date:		
Updates Since Last Report	Highlights	
	Issues and Barriers	
	Key Upcoming Events	
	Team Headline	
	Financial Headline	

Check for Accountability. After business leadership teams have agreed to the final Six Sigma project list, project accountability is then assigned to the owning manager. The owning manager (sometimes called the Project Champion) reviews the projects with his or her business segment team to validate alignment with the needs, strategy, and financial targets of the business.

Based on approval by the business segment team, the owning manager develops a project charter for each selected project with clearly defined objectives, timing, and linkage to the business. The output of this step is a complete project list, complete with project charters.

At the business level, each chartered project must have a Champion who is held accountable for the success of the Black Belt, and the project Business Management (Leadership Team) must also be held accountable for the success of key projects to ensure they are resourced correctly and given the priority for completion. A senior executive with one of our clients stated, "There are no unsuccessful Black Belts, only unsuccessful Champions."

At the personal level, each team member, Black Belt or Green Belt, Project Champion, and Leadership Team member must be held personally accountable for the success of the project. Typically, this is reflected in performance management planning. When considering the myriad of actions necessary to deploy Six Sigma, establishing and tracking accountability for the Six Sigma projects is the requirement of long-term success and preventing Six Sigma from turning into the program of the month. Someone needs to know they might be in some trouble if their project fails.

Review Six Sigma Projects. To ingrain Six Sigma into your culture, set the expectation that the Six Sigma program and projects will each be reviewed systematically and at the project owner level and a higher level. The projects reviews should rigorously address these issues:

- Clearly connected to business priorities:
  - Strategic and annual operating plans.
- Reasonable scope—doable in four to six months:
  - Project support often decreases after six months.
  - Scope too large is a common problem.
- Importance to the organization is clear:
  - People will support a project when they understand the importance.
- Has support and approval of management:
  - Needed to get the resources and remove barriers.

Figure 9.17 shows a typical project summary report that relates project progress to closing the gap between process baseline and process entitlement. It is easy to follow and easy to check project status. This chart is directly derived from the metrics chart in the project charter.

Typical Project Summary

Metric	Baseline	NOW	Goal	Entitlement	Units
Yield	50%	60%	80%	98%	%
COPQ	520K	320K	120K	60K	$/A
Rate	1.4M	1.6M	2.2M	3.0M	Pairs/A

Figure 9.17 The short form of a project review status report. This report includes current performance on three metrics, the goal, and the entitlement.

At a larger scale, Figure 9.18 shows summaries of Six Sigma metrics and how their projects are working by division. The right column, Business Impact, is the most important column for driving culture change. If the projects are effective, so is the program. This column indicates that this $4 billion business has achieved over $50 million in financial results in three quarters. That's $50 million that wasn't in the original annual operating plan.

Example of Using Metrics to Drive the Program
(Actual Q3 Metrics for a $4B Business)

SBU	Rolled Throughput Yield (%)			Cost of Poor Quality (K$)			Capacity (Ratio)			Business Impact Real Dollars
	Baseline	Goal	Now	Baseline	Goal	Now	Baseline	Goal	Now	YDT
Division A	72.6	82.5	76.8	81,843	48,828	27,567	1.00	1.31	1.48	$23.7M
Division B	81.3	85.4	86.5	255,779	200,935	73,516	1.00	1.10	1.16	$14.2M
Division C	83.3	85.7	87.7	334,566	31,936	24,629	1.00	1.06	1.06	$4.7M
Division D	86.3	92.6	91.5	37,800	25,500	23,225	1.00	1.20	1.21	$4.6M
Division E	63.7	70.5	70.0	111,167	73,196	64,479	1.00	1.21	1.22	$5.1M
Company Summary	78.1	82.5	80.8	539,023	394,891	215,440	1.00	1.14	1.22	$52.3M

Figure 9.18 This is a quarterly report—division by division—of the comprehensive impact that Six Sigma projects have on overall performance. The actual financial benefits are listed in the last column.

Finally, Figure 9.19 shows a project review schedule for different leadership levels of a company. At the highest level, project and program ought to be reviewed at least quarterly. At the Project Champion level, projects should be reviewed at least once per week to keep the project on track.

Project Review Schedule

Org Level	Quarterly	Monthly	Bi-Monthly	Weekly
Company Leadership	X			
Division/Functional Staff	X			
Plant Staff		X		
Middle Managers		X		
Six Sigma Project Champions				X

Figure 9.19 A recommended project or Six Sigma program review at different levels of leadership in a company.

SUMMARY OF SIX SIGMA PROJECT SELECTION

Why Projects Matter. Why does selecting great projects for your initial Six Sigma training really matter? High project success rates give you faster acceptance of Six Sigma as a program. Companies successful in deploying Six Sigma develop a pay-as-you-go benefit stream. Benefits start immediately with each training session.

Also, by achieving early results, you will optimize your overall return on training investment. You provide momentum for Six Sigma and you take the wind out of the sails of the naysayers. For example, within the first six months, the engineered materials sector of AlliedSignal chalked up some $50+ million in real financial benefits. There was no problem moving the Six Sigma program forward for a long time.

Why Projects Are Unsuccessful. There are at least four reasons for unsuccessful projects. First, the business need for the project has not been clearly established. This was okay in the TQM days, but not with Six Sigma. When Six Sigma teams know they are working on an important problem for the business, motivation is not an issue.

Second, the Six Sigma team, especially the Black Belt, has not been allocated adequate time to work on the problem. This problem ties into the business need. If an organization knows there is a $1 million project benefit in six months, providing adequate time is not a problem.

Third, the leadership accountability for the project is not clear. No one is personally on the hook for results. By clarifying accountability, you may be able to run, but you can't hide.

And fourth, the scope of the projects is either too narrow or too wide. The project is either small, not strategically important, or not big enough to demand the need for multiple teams to succeed. Projects need to be scoped properly and scoped in such a way that it can be completed in 6 to 12 months. To improve our success rates, we want to cover the following issues:

1. **Clear Business Need**—Direct link to critical top-line or bottom-line business issue. Clear business benefits.
2. **Clear Support**—Highly motivated Champions and Black Belts. Good visibility.
3. **Clear Accountability**—Personal incentives and potential jeopardy to operating plan.
4. **Clear Project Scope**—Significant impact within an explicit, limited time frame.
5. **Clear Applicability**—Good fit with Six Sigma tools and methodology.

How Good Is Your Project Selection Process? Like any process, there should be some metrics to measure to ensure the process is working right. In Chapter 14, I talk about software that allows a company to track projects and obtain some of the following metrics:

- Percent of projects completed in four to six months.
- Average $$ savings per project.
- Average percent benefits actually realized.
- Percent of projects tackling operating plan barriers.
- Percent of projects generating top-line growth.
- Retention rate in Black Belts and Green Belts.
- Percent of Black Belts and Green Belts certified.
- Project completion rate.
- Percent of projects included in the annual operating plan.

Accountability is the key word in project selection. Developing a system by which projects are prioritized, selected, and resourced is an important step in your Six Sigma deployment. Good luck and happy hunting!

10 Creating Six Sigma Executive and Leadership Workshops

With Dan Kutz

Every Six Sigma deployment starts with a series of leadership workshops to create readiness. Because the deployment of Six Sigma and technical nature of Six Sigma are unique in many ways, the company's leadership must be aligned early. The company must understand Six Sigma, and define the program expectations and the lines of accountability.

These early sessions encompass a significant amount of training but also include the workshop format where real deployment work gets done. These are the specific workshops I will address in this chapter. Each workshop discussion will include the length of the workshop, what the attendee should know, and what the attendee should do:

- Executive Team Workshop
- Business Team Workshop
- Deployment and Project Champion Workshop
- Support Organization Workshops (Finance and Human Resources)

This chapter will also discuss a customization process and being successful and avoiding problems.

EXECUTIVE TEAM SIX SIGMA WORKSHOP

The Executive Team Workshop is the first Six Sigma training and work session delivered in a deployment. Assuming you have selected an external consulting group to support the deployment, before the session is designed, these consultants perform initial interviews of each member of the executive team. The interview team is one or two consultants plus the Six Sigma Deployment Champion within the company.

The purpose of these interviews is to assess the base knowledge of the Executive Team in Six Sigma, the company history of process improvement, and what each member views as the important strategic issues to the company. Some initial bullets addressing the purpose of the interviews from the consultants' point of view are the following:

- **Purpose**
  - Individual introductions and interviews with staff and line Senior Executives of a new client.
  - Develop a basic understanding of the critical goals, issues, and culture of the organization.
  - Provide an opportunity for preliminary questions and answers on implementation and technical matters.

Later in this chapter, I will present actual examples of output of the interviews upon which the Executive Workshop was based. This information was the first agenda item for this workshop so the presenters know if they're on target or off base. Following is a list of failure modes that were reported in the interviews for previously unsuccessful initiatives:

- Lack of individual accountability and consequences
- Staff driven
- Many unconnected initiatives
- Lack of structured methodology
- Lack of management reviews and assessments
- Lack of clear measures, goals, and external benchmarks
- Lack of sharing among units
- Lack of clear expectations
- Lack of verification of benefits
- Difficulty in sustaining the gains

These points were very helpful in designing the Executive Workshop. The comments provided a blueprint for the agenda. And we could use these points as an assessment for the workshop at the end. Did we cover these issues or not?

Executive Team Workshop Length. The length of the Executive Team Workshop has varied from one day to three or four days. Considering the amount of information to be conveyed and the action planning to be done, I recommend at least two days for this workshop. The basis for this recommendation is that many members of the Executive Team are new to Six Sigma.

I have found that during one evening (the evening of the first day), a lot of thinking is done. Great questions and comments show up at the beginning of the second day. A one-day session doesn't allow the time to consider the effort. Now, let's look at the key deliverables: what the Executives should know and what they should be able to do.

- **Key Deliverables**
  - What Executives should know:
    - Six Sigma introduction
    - Deployment case studies
    - Lessons learned and failure modes
    - Project selection methodology
    - Project examples
    - HR requirement
    - Executive role in the program
    - What leading the effort looks like
    - Roles and expectations
  - What Executives should be able to do:
    - Engage in leading the effort
    - Provide input to the program
    - Provide input to the creation of a deployment plan or feedback on a draft deployment plan
    - Set financial targets
    - Assignment and responsibility of financial targets
    - Order of deployment by business and functional area
    - Scope of the Six Sigma effort
    - Understand the initial steps involved in implementing Six Sigma in their organizations

The next sections provide you with a generic outline for the Executive Team Workshop and an actual example of a three-day workshop for a very large company. These workshops have had relatively small attendance consisting of the CEO's direct reports—maybe 12–15 people. They've also had a very large attendance, anywhere from 60 to 150 people. The workshop customization process will help the company determine the goals of each workshop and the target attendees. The customization process will be presented later in this chapter.

A GENERIC OUTLINE FOR THE EXECUTIVE TEAM WORKSHOP

- **Purpose**
  - Provides overview of Six Sigma and the leadership roadmap to institutionalize the program
  - Provides leaders understanding of the steps to major culture change
- **Potential Audience**
  - Executive Team, Division Heads, Business Heads, Operations leadership teams, Functional Heads
- **Course Content**
  - Six Sigma and other initiatives
  - Six Sigma overview
  - Selecting the right projects
  - Selecting and training the right people
  - Breakthrough roadmap and tools
  - Leading for excellence
  - Sustaining the gains
- **Duration:** One to four days
- **Deliverables**
  - High-level awareness of Six Sigma and the leadership roadmap
  - Defined expectations and accountabilities
  - Targeted financial targets and Critical Ys
  - High-level deployment plan
  - Identification of Champions
  - Initial list of projects and Black Belts

ACTUAL THREE-DAY AGENDA

- **Day One**
  - Introduction (8:00–8:15)
  - Six Sigma at Company X (8:15–9:00); Company X CEO
  - Break (9:00–9:15)
  - Introduction (9:15–10:00)
  - Six Sigma overview (10:00–11:00)
  - Lunch (11:00–1:00)
  - Six Sigma overview continued (1:00–1:30)
  - Six Sigma tools (DMAIC)
    - Introduction (1:30–2:00)
    - Define Phase (2:00–2:15)
    - Break (2:15–2:30)
    - Measure Phase (2:30–4:00)
    - Break (4:00–4:15)
    - Measure Phase continued (4:15–5:00)
  - Reception (5:00)
- **Day Two**
  - Six Sigma tools continued
    - Analyze/Improve Phases (8:00–9:45)
    - Break (9:45–10:00)
    - Control Phase (10:00–10:30)
    - Case studies (10:30–11:00)
    - Lunch (11:00–1:00)
  - Project identification and selection (1:00–2:30)
  - Break (2:30–2:45)
  - Project identification continued (2:45–3:15)
  - Lessons learned (3:15–3:45)
  - Break (3:45–4:00)
  - Driving Six Sigma to the bottom line (4:00–4:30)

  - Wrap-up/elevator speeches (4:30–5:00)
  - Reception (5:00)
- **Day Three**
  - Leading Six Sigma (8:00–8:30)
  - OPEN for Q&A
  - Break (9:30–9:45)
  - Deployment (9:45–10:30)
  - Leadership expectations (10:30–11:00); Company X Leader
  - Lunch
  - Breakouts (1:00–?)
  - Operations Committee panel (?–4:30)
  - Breakout reports
  - Q&A
  - Wrap-up (4:30–5:00) CEO

Business Team Six Sigma Workshop

The Business Team Workshop is similar to the Executive Team Workshop. The underlying difference between the two workshops is the focus. The Executive Team Workshop focuses on deploying Six Sigma across the entire company, whereas the Business Team Workshop focuses on deploying Six Sigma within a specific business.

The Business Team Workshop allows each business the decentralism to customize their Six Sigma program to their specific and unique needs. This customization is done within constraints of the deployment plan approved by the Executive Team and the Six Sigma Steering Team, which is lead by the Initiative Champion.

An example of this approach is AlliedSignal when launching Six Sigma in 1995. AlliedSignal had three business sectors: the aerospace sector, the automotive sector, and the engineered materials sector. These sectors had radically different market environments, and it made sense to allow the sectors a little freedom in deploying Six Sigma effectively. The deployment had a great deal of central input to ensure consistency where it made sense and, likewise, customization where it made sense.

For example, all three sectors simultaneously launched two waves of Black Belts each in January 1995. AlliedSignal, as a company, launched a total of six Black Belt waves simultaneously. However, the Black Belt curriculum was slightly different for each sector to reflect the difference in manufacturing methodology. Manufacturing a jet engine (aerospace) is radically different from manufacturing a spark plug (automotive) or manufacturing an advanced polymer chemical (engineered materials).

- **Key Deliverables for Business Team Workshop**
  - Workshop length:
    - About the same as the Executive Team Workshop
    - One to four days
  - What Business Team should know:
    - Deployment plan for the company (presented by Business Team leader)
    - Goals for the business team ($ results, # of Belts and projects, linkage to existing targets)
    - Six Sigma intro, project selection methodology, example projects, and HR requirements
    - Understand their role in the program
  - What Business Team should be able to do:
    - Provide financial targets, Critical Ys, and project clusters
    - Effectively select projects
    - Identify Belt candidates
    - Provide input to the business' deployment plan
    - Clarify accountability for results

A GENERIC OUTLINE FOR THE BUSINESS TEAM WORKSHOP

- **Purpose**
  - Chartered by the Executive Team
  - Provides overview of Six Sigma and the leadership roadmap to institutionalize the program
  - Provides understanding of the steps to major culture change
  - Provides ability to effectively review and drive the program and breakthrough projects
- **Audience**
  - Division, Business, Operations leadership teams

- **Course Content**
  - Six Sigma overview
  - Selecting the right projects
  - Selecting and training the right people
  - Breakthrough roadmap and tools
  - Managing for excellence
  - Sustaining the gains
- **Duration:** Four days
- **Deliverables**
  - High-level deployment plan
  - Financial targets and Six Sigma expectations
  - Clear roles and responsibilities for Six Sigma Leaders, Champions, Black Belts, and Master Black Belts
  - Clarified deployment plan with a firm schedule of actions to be taken
  - Guidance on project selection and prioritization
  - Six Sigma reporting structure

ONE-DAY PURPOSE AND AGENDA—SAMPLE OF AN ACTUAL WORKSHOP

- **Purpose of the Workshop**
  - To gain an understanding of the kinds of business problems Six Sigma can solve
  - Clearly establish the important 2004 project clusters for the business
  - To identify and select 15 to 30 projects aligned with key business goals for 2004
  - Clarify and understand the roles of business leaders
  - Next steps
- **Agenda**
  - 08:20 Group expectations/survey results (8:20–8:45)
  - 09:00 Agenda, introductions, and expectations
  - 09:30 Introduction to Six Sigma—leadership perspective
    - What is Six Sigma?
    - Why should we deploy Six Sigma?
    - Results expectations

- 10:15 Deployment case examples
    - Example deployment #1 and lessons learned
    - Example deployment #2 and lessons learned
- 10:00 key measures and project clusters
- 11:15 Project selection methodology and project charters
- 12:00 Lunch
- 1:00 Process improvement methodology for operations
- Manufacturing application
    - Manufacturing project example
    - Transactional application
    - Transactional project example
- 3:00 Design for Six Sigma introduction
    - Design project example
- 4:00 Deployment strawman for company
- 4:30 Summary and feedback
- 5:00 Adjourn

Actual Four-Day Business Team Agenda

- **Day One**
    - Process Excellence overview
    - Six Sigma overview
    - Balanced scorecard and metrics
    - Project selection and scoping
- **Day Two**
    - Balanced scorecard and metrics
    - Project selection and chartering
    - Process mapping
    - Cause and effects matrix
    - Black Belt case study
    - Measurement systems study

    - Definition of terms
    - Questions to ask
- **Day Three**
  - Capability study
    - Shape, center, and spread of data
    - Questions to ask
  - Failure Modes and Effects Analysis
    - Overview/background
    - Questions to ask
  - Black Belt case study
  - Multi-vari study
    - Philosophy
    - Inputs versus outputs
    - Questions to ask
  - Experimental design
    - Factorials and screening studies
    - Questions to ask
  - Control plans
    - Components and systems
    - Questions to ask
  - Black Belt case study
  - Deployment planning
    - Lessons learned
    - Role of the Champion and Master Black Belt
  - Exercise continues
- **Day Four**
  - Deployment planning continued
    - Review goals and metrics
    - Process for program review
    - Training deployment
    - Communications plan
    - Reward and recognition

- Vision exercise
- Elevator speeches
- Wrap-up, and pluses and deltas

The Business Team Six Sigma Workshop holds importance because this is the venue for the businesses to take ownership and accountability of Six Sigma. After a series of Business Team Workshops, the Six Sigma program is no longer strictly a corporate-level program but a business-level program.

These workshops allow business leadership to experience the Six Sigma tools, terminology, and methodology. This is also the opportunity to commit to disciplined project identification and prioritization. The businesses will develop a potential project list for their businesses with the following:

- Business impact
- Resource requirements

Most importantly, the businesses will define the deployment plan for Six Sigma in their businesses. Each business will follow Kotter's eight stages to culture change, and there will be a subtle competition among businesses for the best Six Sigma results.

DEPLOYMENT AND PROJECT CHAMPION WORKSHOPS

The objectives of Champion Workshops are to ensure that the Champions understand their company's Six Sigma program and their role in that program. Specifically, as the result of this training, Champions will understand what Six Sigma is, and how it differs from other quality/cost reduction/customer satisfaction programs. They will know how to deploy a successful Six Sigma program. In addition, they will know how to identify, scope, prioritize, and charter Six Sigma projects. They will also know how to select Black Belts and Green Belts.

Along with all that, the Champions will understand how to review projects and provide coaching for their Black Belts and Green Belts. Finally, in learning the guts of Six Sigma, the Champions will gain enough knowledge of the Six Sigma tools so they know what questions to ask during project reviews. Going from identifying and chartering high-impact projects to the successful completion of these projects by their Black Belts and Green Belts is the ultimate intent of the Workshop.

- Key Deliverables for Deployment and Project Champion Workshops
  - Workshop length:
    - Two to five days in length
  - What Champions should know:
    - Deployment plan for the company (presented by Initiative Champion)
    - Six Sigma concepts
    - Six Sigma tools and roadmap basics
    - Identify, scope, prioritize, and charter Six Sigma projects
    - Nontechnical project review
    - Lessons learned and failure modes
    - HR requirement
    - Black Belt and Green Belt selection
    - Understand their role in the program
  - What Champions should be able to do:
    - Deploy a successful Six Sigma program
    - Project selection methodology and project chartering
    - Identify Black Belt and Green Belt candidates
    - Provide nontechnical project mentorship
    - Remove barriers to project success

A GENERIC OUTLINE FOR THE DEPLOYMENT AND PROJECT CHAMPION WORKSHOPS

- **Purpose**
  - Provides overview of Six Sigma and the leadership roadmap to institutionalize the program
  - Define project selection techniques
  - Review/refine project charters for Black Belt projects
  - Determine communications methods for Champion networking
  - Develop reporting methods/requirements for Black Belts and breakthrough projects

- **Audience**
  - Six Sigma Deployment Champions, Six Sigma Project Champions
- **Workshop Content**
  - Six Sigma overview
  - Project selection and definition
  - Breakthrough roadmap and tools
  - Planning for communication, reporting formats, project reviews, and Champion networking
- **Duration:** Two to three days
- **Deliverables**
  - Detailed deployment plan
  - Clear roles and responsibilities for Champions, Black Belts, and Master Black Belts
  - Commitment to select high-impact projects

ACTUAL EXAMPLE OF THE DEPLOYMENT AND PROJECT CHAMPION WORKSHOPS

- **Day One**
  - Introduction (8:00–8:30)
  - Executive interview results (8:30–8:45)
  - Six Sigma overview (8:45–9:30)
  - Break (9:30–9:45)
  - Six Sigma overview (9:45–12:00)
  - Six Sigma card drop (exercise that demonstrates Six Sigma)
  - Lunch (12:00–1:00)
  - Project reviews (1:00–1:30)
  - Team discussion
  - Six Sigma tools (1:30–3:00)
    - Define Phase
    - Measure Phase
  - Break (3:00–3:15)

- Six Sigma tools continued (3:15–5:00)
    - Analyze Phase
    - Improve Phase
- **Day Two**
    - Six Sigma tools continued (8:00–8:45)
        - Control Phase
    - Lean Design
    - Six Sigma tools continued (8:45–9:45)
        - Lean Sigma
        - Transactional Six Sigma
    - Break (9:45–10:00)
    - Project reviews (10:00–10:30)
    - Team discussion
    - Project selection and prioritization (10:30–12:00)
    - Lunch (12:00–1:00)
    - Project selection and prioritization (1:00–3:00)
    - Break (3:00–3:15)
    - Project selection and prioritization (3:15–4:00)
    - Project chartering (4:00–5:00)
- **Day Three**
    - Six Sigma roles and responsibilities (8:00–8:45)
    - Managing Six Sigma and sustaining the gains (8:45–9:30)
    - Break (9:30–9:45)
    - Deployment planning (9:45–10:45)
    - Deployment cases (10:45–11:15)
    - Lessons learned (11:15–11:45)
    - Elevator speeches (11:45–12:00)

Training the Six Sigma Champions is a special challenge because of their complex role within Six Sigma. These resources have to learn the technology of Six Sigma so they can be effective project reviewers. In my sector of AlliedSignal, the Champions (and my boss) attended the entire Black Belt training during the first Black Belt training. I recommend your Champions doing this, but I have seen few companies follow that path.

The Champions are also accountable for results and must be trained in leading and managing Six Sigma. As old Champions move on to other positions and new Champions come on board, the annual training plan should include training and certifying the new Champions.

A PROCESS FOR RAPIDLY CUSTOMIZING EXECUTIVE/ CHAMPION WORKSHOPS

Due to the difficulty of arranging time for the Executive Team, Business Teams, and Champion training sessions, the importance of knowing the audience, knowing the business environment, and being on target with the workshop should not be underestimated. Some good front-end work will ensure success here, especially if an external consulting group is supporting the deployment effort.

Executive Interviews. Executive interviews are an excellent way to start a deployment. This is a simple process whereby a large proportion of the Executive Team participates in a 30 to 45 minute face-to-face interview with the Six Sigma Initiative Champion and an external consultant.

An interview guide provides the foundation of the interviews to ensure some level of consistency and prevent the omission of important questions. The following is an example of an interview guide. This example consists of 12 questions that are developed based on the company and the interview participants. There are some additional recommended questions toward the end of this example.

Executive Session Interview Guide

1. On a scale from 1 to 10, how would you rate your personal experience in deploying Six Sigma?

   1—I have very little knowledge; I've heard of it.

   3—I've read about it and had some training in it.

   5—I've participated in a deployment once or twice.

   7—I've lead a deployment.

   10—I am very comfortable with all aspects of leading a deployment.

2. On a scale from 1 to 10, how would you rate your personal knowledge of Six Sigma methodology and tools?

   1—I have very little knowledge; I've heard of it.

   3—I've read about it and had some training in it.

5—I have received Black Belt or Green Belt training.

7—I successfully completed my certification project.

10—I successfully completed several Six Sigma projects.

3. As you know, we will be holding a ____-day Six Sigma Executive/Champion training session on ______ date. The focus of this session is to provide a general understanding of the Six Sigma process and to identify potential projects for our improvement effort. Do you have any additional areas of interest that you would like to see addressed in this session?
4. Please review the topics listed below. Which of these do you consider to be a "must" to review in the Champions Workshop? Are there any topics listed below that do not need to be reviewed in the Workshop?
   - An overview of the Six Sigma processes:
     i. DMAIC (Operations Six Sigma)
     ii. DFSS (Six Sigma for R&D)
     iii. Transactional Six Sigma
     iv. Lean Sigma
   - An overview of Six Sigma methods and tools
   - An overview of the process for deploying Six Sigma
   - An overview in how to conduct a good project review
   - Some cases describing other deployments
5. What would you consider to be the three highest-priority issues facing your business today? What issues would you like us to address as we help you deploy Six Sigma training throughout your organization?
6. What are your key performance metrics (big Ys)? How frequently are these metrics reported and used by your organization?
7. What areas of your business would you say could benefit most from improved execution?
8. What would be a rough estimate for the cost of inefficient execution (cost of poor quality) in the following areas?

   Operations

   Marketing

   Transactional processes

   New product commercialization

9. How well connected are current operations, marketing, and technology programs to your overall business strategy and goals?
10. What role do you see for new product commercialization in meeting your business objectives?
11. Are benchmark and customer satisfaction studies performed to determine best in class for your key products, services, and administrative functions?
12. Where are current improvement areas for reducing waste and improving process efficiency?

As companies prepare to deploy Six Sigma, a key aspect is project selection. Consider the following when selecting projects:

- Where are the areas that contribute to excessive costs?
- Where are the areas that cause excessive time delays?
- What processes are contributing to poor process efficiency? Think about all work processes, not just manufacturing.
- What are your customer's major complaints relative to your products and services?
- What are your key business metrics and which of these exhibit the largest performance gap?
- What are the 3–5 key top-down business issues?
- What are the key issues from a bottom-up view (how many times result in quick hits or early successes)?

Executive Interview Report. Once the interviews are completed, the results become part of the leadership workshop. The effectiveness of this stems from the fact that the company's current situation and strategy were considered when designing the workshop.

Presenting the interview results provides an opportunity to get the participants on the same page with respect to the company's business issues. The interview results provide insights into what the opportunities are and where Six Sigma can help. An actual example of how the information gained from the interviews was put together as a workshop opening.

There are two classes of information gleaning from the interviews and for the workshop. The first class of information has to do with the company's business issues and improvement opportunities. The second class of information has to do with the base knowledge each interviewee has of Six Sigma. This information, shared textually or graphically as seen next, is a great way to start a workshop.

Q1. WHAT ARE THE THREE HIGHEST-PRIORITY ISSUES FACING YOUR BUSINESS TODAY?

- Responses
  - **Respondent #1:**
    - Proper integration of recently acquired businesses.
    - Communicating the 2–5 year vision for our company to the broader Executive management team.
    - Sarbanes-Oxley compliance with more than 20 different A/R systems.
    - Huge opportunity in working capital (inventory).
  - **Respondent #2:**
    - Identifying the project Champions and managers.
      - Our company has more than 100 IT people working in more than 10 sites. Need to retool resources from infrastructure to applications side.
    - Data availability and data quality for internal management of the business.

Q2. WHAT IS YOUR PERSONAL EXPERIENCE IN DEPLOYING LEAN SIX SIGMA?

X				
X	X	X		X
X	X			X
X	X	X	X	X
1	3	5	7	10
Very little, but have heard of it	Read about it and have some training	Participated in a deployment	I've led a deployment	Very comfortable with all aspects of leading deployment

Figure 10.1 A graph depicting how the interviewees responded to their experience in deploying Six Sigma on a one-to-ten scale. This shows that over half the potential respondents have no experience in deployment.

Q3. WHAT IS YOUR PERSONAL KNOWLEDGE OF SIX SIGMA METHODOLOGY AND TOOLS?

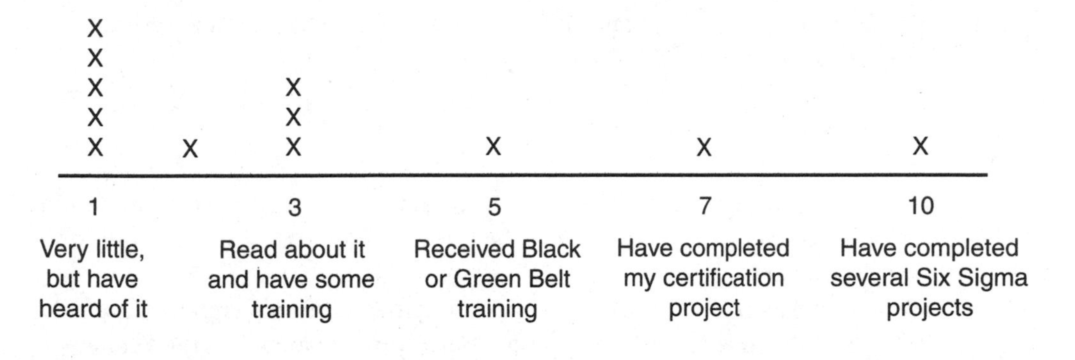

Figure 10.2 A graph depicting how the interviewees responded to their knowledge of Six Sigma methodology and tools. This shows that over half the potential respondents have little to no knowledge of Six Sigma tools.

Q4. WHAT WOULD YOU PERSONALLY LIKE TO GET OUT OF THIS SESSION?

- **Responses**
  - **Respondent #1**
    - What are the next steps?
    - Understanding that this is a journey; a deployment process.
    - A focused list of projects and the associated returns.
    - The project Champion; the project leader.
    - Agreement that we will drive the projects.
    - Determine the *realistic* opportunity of what we can get done.
    - A clear understanding and agreement as to what we all expect to get out of this.
    - With our CEO's input and priorities, what are we going to work on?
    - Avoid trying to tackle 30 items.
  - **Respondent #2**
    - A deeper understanding of the Six Sigma process.

- What the Lean animal is, and how it is defined (versus my Continuous Quality Improvement experience).
- A clear definition of what Lean and Six Sigma means for our company.
- The structure and understanding of the resources to be devoted to these projects to be successful.
- I would like the group to have the realization that this is something we really need to do.

By completing a straightforward interview process, we now have a great start on designing the leadership workshops. We know what the business issues are and what the concerns about Six Sigma are.

Now it's time to develop the agenda and supporting materials and exercises. For example, we can use the interviews to assess how much time to spend on the Six Sigma tools. This is a technical module, and as much as six hours can easily be spent on the tools. Next are the conclusions that were arrived at concerning the Six Sigma tool module in the various leadership workshops:

- **Example: Tools Training**
  - **Goals for Executives and Business Teams**—What are the tools, how do they fit together, how do they drive business benefit? (2 hours)
  - **Goals for Champions**—Overview (so they can understand enough to select projects): What are the tools, how do they fit together, how do they drive business benefit, and what are some questions I can ask to spark discussion with Belts? (4 hours)
  - **Goals for Champions**—In Depth (so they can be better project reviewers): What are the tools, how do they fit together, and how do they drive business benefit? How do I use the tool (classroom application experience), and what are some questions I can ask to spark discussion with Belts? (6 hours)

MATERIALS REVIEW WITH KEY STAKEHOLDERS

Again, for the leadership workshop to be on target, the Six Sigma stakeholders and the consultants review the materials used for the workshop. As with everything else in Six Sigma, the material review is a process. The steps to the process are listed next. The rapid materials review process for fast launch is detailed next:

1. Assemble key representatives (Initiative Champion, subject matter experts, keepers of company values and culture) for four- to eight-hour materials reviews (Exec/Business Team/Champ and/or GB/BB material sets, etc.).
2. Review terminology (session naming, role names, acronyms, etc.).
3. Review look and feel of materials (branding, logos).
4. Review course-level deliverables and roles (week by week if course spans several sessions).
5. Review agenda level with agenda that shows topics, durations, and order.
6. Review module-level learning objectives and key outputs or decisions in each module.
7. Assign subject matter experts to review detailed module content if needed. Fast launch will dictate that detailed module-by-module review is impractical, so limit detail review to crucial content only.
8. Use Kaizen/do-it-right-now work rules when editing materials. To avoid white spaces in time, use two- or three-person teams, co-located to assemble materials, and have materials reviewed in parallel (avoid batching).

I followed this process when working with a large company. We even reviewed some of the materials with the new CEO. A large group of subject matter experts met with the consulting experts to finalize the training material. This activity resulted in a three-day workshop that ended with a standing ovation. Every event has the potential of being this successful with the right planning.

BEING SUCCESSFUL, AVOIDING PROBLEMS

Because the leadership workshops occur at the beginning of a Six Sigma deployment, their success is crucial when launching the deployment of the new initiative. The workshops provide the opportunity to build rapport among the leaders of the company and to know in advance the team to be involved and their issues. With the right customization, many of the issues will be surfaced safely and addressed assertively. The final deployment plan will be well known with a high level of enthusiasm.

Order of the Workshops. By developing the content of the leadership workshops after the interviews, the order of the workshops may also be established. Although the general order is (1) Executive Team Workshop, (2) Business Team Workshops, and (3) Champion Workshops, this does not have to be your order. You can order the workshops in a way that works best for your company and your Six Sigma deployment.

Workshop Leaders. For companies using external consulting support, the presenters are usually a combination of consultants and company employees. Who the company selects to lead the workshop gives the participants a sense of the importance of the program. Companies sometimes allow their newly hired Master Black Belt to do the session.

Unless this person comes across as a business leader, he or she will be viewed as a technical resource that doesn't know much about running a business. Assigning the workshop leader slot to a senior, highly respected leader is the way to go. This person should be a leading contender for the Initiative Champion position.

Timing of the Deployment. The successful leadership workshop will positively affect the timing for the deployment. The deployment speed, rather than being conservatively slow to avoid risk, will be done at light speed because leadership will understand the benefits of a fast start. For example, at 3M, Jim McNerney launched the first wave of Black Belts only three weeks after the Executive Team Workshop. The launch was highly successful.

Financial Targets for Six Sigma. Most importantly, the leadership workshops have the advantage of deploying the expected financial targets for the program well in advance of the first Black Belt and Green Belt training even beginning. These clear targets will directly affect each business's, function's, or site's sense of urgency for selecting the right projects. The financial targets also send the message that Six Sigma is not just a quality program—Six Sigma is a business improvement program that belongs to the businesses.

Nerves of the Staff: Expect it. Even before the first leadership workshop, a group of staff members will be involved in putting the leadership workshops together. This group will be very nervous because of their lack of expertise in Six Sigma. They have to work hard to learn the guts of the program and also allow a set of consultants to help make some of the important decisions. They will sometimes be in direct conflict with the consultants because they were involved with similar initiatives previously.

One way to get around that conflict is the way one CEO did it. He told his staff that the company would use methodology that they were licensing from the consultants rather than try to create the program from scratch. This saved the company a lot of time, and the staff members still influenced the program positively.

Other workshops may include a one-day financial workshop and a one-day HR workshop. The agendas for each are shown next.

Example Agenda of a Financial Workshop

- 8:30 Introductions/SPACER
- 9:00 Six Sigma overview
- 10:30 Information sharing
- 11:45 Lunch
- 12:45 Role of finance in Six Sigma—creating a finance and Six Sigma partnership

- 3:15 Finance resource requirements
- 4:30 Next steps/milestones
- 5:00 Adjourn

Example Agenda for Human Resources Workshop

- **Training Session**
  - Provide introduction to Six Sigma and Lean methodologies.
  - Overview of improvement methods.
  - Show potential improvement opportunities in Human Resource areas.
- **Working Session**
  - Work through HR aspects of deployment plan.
  - Improvement program organization structure.
  - Recognition and rewards.
  - Appraisal and performance.
  - Retention and career-path planning.
  - Change management considerations.
  - Communication plan.
  - Team effectiveness.

The quality of the leadership workshops at the beginning of the deployment will drive the spirit in which the program progresses. Doing the right upfront work and getting the right people involved early will ensure that your workshops are successful.

11 Selecting and Training the Right People

An integral stage in driving large-scale change is building the guiding coalition. This is the second stage of John Kotter's eight stages for leading change protocol covered in detail in Chapter 3, "Six Sigma Launch Philosophy." Kotter asserts that the people within the guiding coalitions must have strong positional power, broad expertise, and high credibility.

Extensive training in Six Sigma is important because the guiding team should have the capability to lead Six Sigma and the technical expertise to understand and use Six Sigma. These factors add up to credibility in driving Six Sigma. In addition, a supporting infrastructure for the coalition must exist with well-defined processes. One of the infrastructure processes is guiding ongoing Six Sigma training, because training, if done right, empowers your people to take action via process improvement projects.

Six Sigma deployments present the two challenges of launching new methods of process improvement, new organizational structures, and new positions to lead the program. Training is the success factor in any Six Sigma deployment. Training supports Kotter's stage, "Empowering employees to broad-based action."

The leadership team in AlliedSignal received substantial training in Six Sigma. I recall holding several one-on-one training sessions with the CEO during the early days of deployment. This chapter will present some common training agendas for your review. Chapter 8, "Defining the Six Sigma Infrastructure," when discussing the Six Sigma infrastructure, carefully covered the selection and the roles of the deployment team members plus a summary of the training for each position.

The Six Sigma training accomplishes several actions simultaneously. The training provides students with the technical knowledge of Six Sigma. By training everyone on the deployment team, the team starts learning how to work together as a team. We want to start the deployment with the right people, and training will help determine who might be the right people and provides the right people a common ground upon which to stand. This chapter addresses selecting and training the right people for the right positions. The training we discuss in this chapter includes the following:

- Master Black Belts
- Black Belts
- Green Belts

The training for the following positions

- Initiative, Deployment, and Project Champs
- Financial Resources
- Human Resources

has been covered in Chapters 8 and 10 already and will not be reviewed in this chapter.

Each person in each one of these positions must have leadership capabilities and personal motivation to effectively change the culture. Each deployment team member has a role common with the others. They each must

- Communicate Six Sigma initiative.
- Set aggressive goals and drive implementation.
- Develop and implement the improvement strategy.
- Track and report impact of the change initiative.
- Break down business barriers.
- Create supporting systems.
- Develop performance databases.
- Have internal Six Sigma technical expertise.
- Celebrate successes.

Six Sigma training events are customized for each company deploying Six Sigma. There is not a turnkey set of customized training material. Once the Executive Team has clarified where Six Sigma fits strategically, the training program is designed to meet those expectations. My consulting company has a huge training portfolio, but no client has deployed every course during the deployment. And the courses they did deploy were customized to reflect their company's industry, markets, and strength and weaknesses. Let's take a look at the broad portfolio of Six Sigma training courses.

Six Sigma, Lean Sigma, and Lean Training Portfolio—Enterprise Wide

We will now review a portfolio of training programs for both Six Sigma and Lean both separately and integrated. These programs cover the spectrum from relatively shorter and less technical leadership workshops to lengthy and highly technical "Belt" training and some methodological specialties such as demand forecasting.

The Six Sigma Steering Team has the opportunity to move Six Sigma training literally into every part of the company. But, as in project selection, prioritizing the Six Sigma training targets and goals should be done within the first 90 days and the plan executed accordingly. The bulk of the training will be aimed toward Black Belts and Green Belts. Specialized training aimed at specific functions (e.g., Supply Chain Management) will also be added to the deployment plan when it fits strategically. The seven categories we will investigate include the following:

1. Leadership Workshops
2. Business and operational assessments
3. Value creation and management
4. Operational processes
5. Business (transactional) processes
6. Lean methods
7. Master Black Belt

Training Portfolios: Six Sigma, Lean Sigma, and Lean Training

1. **Leadership Workshops and Participants**
   - Strategic Planning (Executive Team)
   - Executive Overview (Executive Team)

- Business Team Workshop (Business Team)
- Champion Workshop (Initiative, Deployment, and Project Champions)
- Site/Functional Leadership Workshop (Site or Functional Leadership)
- Leadership Green Belt (Any leader)

2. **Business and Operational Assessments—Asset Maximization**
   - Business Assessment
   - Operations Assessment
   - Existing Technology Assessment
   - Transactional/Business Process Assessment
3. **Value Creation and Management: Design for Six Sigma (DFSS)**
   - General SBTI DFSS Overview
   - Design for Six Sigma (DFSS) (Black Belts, Green Belts)
   - Chemical Design for Six Sigma (Black Belts, Green Belts)
   - Voice of the Customer (VOC) (Marketing, Sales Functions)
   - Marketing for Six Sigma (MFSS) (Marketing Function)
   - Technology for Six Sigma (TFSS) (R&D)
   - Six Sigma Process Design (SSPD) (Black Belts, Green Belts)
4. **Operational Processes and Participants**
   - Operational Assessment (Leadership, Champions)
   - Six Sigma in Assembled Products, Chemical Products (Black Belts, Green Belts)
   - Lean Sigma (Black Belts, Green Belts)
   - K-Sigma™ (Black Belts)
   - Analytical GB (Laboratory Green Belts)
   - Asset Dependability (Equipment Reliability Green Belts)
   - Supply Chain GB (Supply Chain Black Belts and Green Belts)
   - Demand Forecasting Workshop (Supply Chain Leadership)
   - Environmental, Health, and Safety (EHS) (EHS Black Belts and Green Belts)
5. **Business (Transactional) Processes and Participants**
   - Transactional Overview (Leadership, Champions)
   - Transactional Green Belt

- Using Minitab, JMP, or XLGB
- Six Sigma Process Design (SSPD) (Functional Leadership, Green Belts)
- Transactional Black Belts
- Transactional Kaizen (Black Belts, Green Belts)

6. **Lean Methods and Participants**
   - Lean—Manufacturing (Black Belts, Green Belts)
   - Lean and Six Sigma (Black Belts, Green Belts)
   - Lean—Transactional (Green Belts)
   - Lean Deployment Approach
   - Kaizen
   - Lean Leader (Lean Leaders)
   - Lean Design (Black Belts, Green Belts)
7. **Master Black Belt**

SIX SIGMA AND LEAN

Many companies deploy Six Sigma and Lean Enterprise initiatives separately. The current trend is to launch both simultaneously to quickly gain the advantages of both technologies. Very simply, Lean addresses business speed and Six Sigma addresses process stability and accuracy. Lean focuses on removing waste in process, leaving only value-added activity. Six Sigma removes variability within any process. Six Sigma and Lean, when integrated, provide the most powerful method of reinventing a business.

WHEN SIX SIGMA?

- Process inputs are variable: suppliers, raw materials, capabilities.
- Process outcomes are not predictable: purity, specs, timing.
- We are creating a new product or process (inventing).
- Quality of internal and/or external products are suspect.

WHEN LEAN?

- Our lead time is too long.
- We are "reacting" to problems.
- Supplier alignment: inventories unstable or slow.
- Customer alignment: impact buys caused by short horizons.
- Inventory levels fluctuate, causing excess overtime, delays.
- Floor space, inventories, flows are not optimal.
- Resources: equipment, people, facilities, cash is not optimal.

By integrating Lean with Six Sigma, we attack multiple business performance gaps simultaneously. The training in both really doesn't take much longer than training for both individually.

LEADERSHIP WORKSHOPS

The leadership training has already been overviewed in Chapter 10, "Creating Six Sigma Executive and Leadership Workshops." These workshops are implemented at the very beginning of Six Sigma. The order of events usually follows this pattern:

- Strategic Planning (Executive Team)
- Executive Overview (Executive Team)
- Business Team Workshop (Business Team)
- Champion Workshop (Initiative, Deployment, and Project Champions)
- Site/Functional Leadership Workshop (Site or Functional Leadership)
- Leadership Green Belt (Initiative, Deployment, and Project Champions)

Recommended Leadership Training

- **Executive Steering Team**
  - Executive Overview
- **Division Leadership**
  - Executive Overview
  - Business Team Workshop

- **Business Leadership**
  - Executive Overview
  - Business Team Overview
  - Full or Leadership Green Belt with Project
- **Site Team Leadership**
  - Site Team Overview
  - Full or Leadership Green Belt with Project
- **Initiative, Deployment, and Project Champions**
  - Champion Training
  - Green Belt or Black Belt with Project

The results of these workshops are the defined Six Sigma infrastructure, trained Six Sigma Champions, a set of prioritized Six Sigma projects, and the planned launch of Six Sigma Black Belt and Green Belt training. The Leadership Green Belt training is a four-day condensed version of Green Belt training. The focus is on the tools and the project review process.

VALUE CREATION AND MANAGEMENT: DESIGN FOR SIX SIGMA

This set of training programs supports the company's efforts to design new products, new technologies, or new customer services. There is a substantial "Voice of the Customer" set of tools involved. Optimally, both technical R&D resources and marketing resources are heavily involved and working together on their new products. Listed next are the training programs along with the generic course length. We will cover one in detail to give you an idea of the course content.

- General SBTI DFSS Overview (One Day)
- Design for Six Sigma (DFSS) (Four- to Five-Week Sessions)
- Chemical Design for Six Sigma (Four- to Five-Week Sessions)
- Voice of the Customer (VOC) (One Week)
- Marketing for Six Sigma (MFSS) (Four Weeks)
- Technology for Six Sigma (TFSS) (Four Weeks)
- Six Sigma Process Design (SSPD) (Two Weeks)

CHEMICAL DESIGN FOR SIX SIGMA—BLACK BELT TRAINING

The first training overview we'll review will be the Chemical Design for Six Sigma (CDFSS) Black Belt program targeting chemical processes and designs. This is an extensive training program consisting of five one-week sessions. The three topic categories include modules on technical design, new product gate review process, and the business of developing new products. Project reviews for each student are completed each week by a team of Project Champions. Each student arrives at the class with a chartered project in hand.

- Week 1
    - Champion/Project Reviews
    - CDFSS Roadmap and Vision
    - Space Tower
    - Stage Gate Process
    - Understanding Customer Value
    - Concept Engineering
    - Market Segmentation
    - Business Planning
    - QFD and C&E Matrix
    - Concept Selection
    - Project Selection/Business Planning
    - Project Management: Principles
    - Product Stewardship/HSE&R
    - Product/Process Development Roadmap
    - Mapping/C&E Matrix
    - Minitab

- Week 2
    - Champion /Project Reviews
    - Design FMEA/Process FMEA
    - Design for Manufacture
    - Defining Statistical Tolerances

  - Basic Statistics
  - Capability
  - Measurement Systems Analysis
  - Presentation Basics
  - Multi-Vari Studies
  - Correlation
  - Manufacturing Cost Estimation
  - Teaming Skills
  - Project Management: Project Mapping

- Week 3
  - Champion/Project Reviews
  - Design FMEA
  - Modular Design
  - Inferential Statistics
  - Design of Experiments
  - Multiple Regression
  - Capital Cost Estimation
  - Advanced Statistics
  - Single Factor Experiments
  - Project Management: Software
  - Full Factorials
  - 2K Factorials

- Week 4
  - Champion/Project Reviews
  - Experiments and Design Tolerances
  - Robust Design Techniques
  - Design Optimization
  - Financial Analysis
  - Fractional Factorial Experiments

- Response Surface Methods
- Advanced Measurement Systems
- Gate Reviews

- Week 5
  - Champion/Project Reviews
  - CDFSS Roadmap and Metrics
  - Reliability Prediction
  - Design and Process Validation
  - Control Methods
  - Documentation
  - Hand-Off Procedures
  - Mixtures/Formulations
  - Multiple Regression
  - Proprietary Positioning
  - Creativity and Brainstorming
  - Product Launch
  - Maintenance and Reliability in Design
  - Process Modeling and Simulation/Advanced Process Control
  - Robust Design/DFM
  - Scale-Up/Dimensional Analysis
  - Control Plans

OPERATIONAL BLACK BELT—MANUFACTURING PROCESS IMPROVEMENT

The next training overview we'll discuss is the Operational Black Belt program targeting manufacturing processes and designs. This is an extensive training program consisting of four one-week sessions. The three topic categories include modules on technical process improvement, effective presentation skills, and team skills. Project reviews for each student are completed each week by a team of Project Champions. Each student arrives at the class with a chartered project in hand.

- **Week One**
    - Champions/Project Review
    - Introduction to Six Sigma
    - Project Chartering
    - Process Mapping
    - Cause and Effects Matrix
    - FMEA
    - Introduction to Minitab/JMP
    - Basic Statistics
    - Basic Quality Tools
    - Measurement Systems Studies
    - Capability Studies
    - Multi-Vari Planning
    - Sampling Considerations
    - Team Formation
    - Introduction to Control Plans
    - Introduction to SPC

- **Week Two**
    - Champions/Project Review
    - Multi-Vari Analysis
    - Confidence Intervals
    - Central Limit Theorem
    - Hypothesis Testing
    - T-Tests
    - One-Way ANOVA
    - Chi-Square
    - Correlation-Regression
    - Introduction to DOE
    - Helicopter Competition
    - Week 2 Review and Action Planning

- **Week Three**
  - Champions/Project Review
  - Full Fractional DOE
  - Fractional Factorial DOE
  - Center Points and Blocking
  - EVOP
  - DOE Planning
  - Capability II
  - Multi-Vari II
  - Multiple Regression
  - Logistic Regression
  - Summary

- **Week Four**
  - Champions/Project Review
  - Attribute Response DOE
  - Autocorrelation
  - Advanced Designs
  - Statistical Control Methods
  - Control Plans
  - Sequential Experiments
  - Sample Size Calculation
  - Response Surface Methods

K-SIGMA BLACK BELT (SIX SIGMA AND LEAN)—MANUFACTURING PROCESS IMPROVEMENT

The next training overview we'll review will be the K-Sigma Black Belt program targeting manufacturing processes and designs. The K stands for "Kaizen" and represents Lean Enterprise technology. This is an extensive training program consisting of five one-week sessions. This Black Belt course is longer than the traditional Black Belt course because it includes training modules in both Six Sigma and Lean.

The three topic categories include modules on technical process improvement for both Six Sigma and Lean, effective presentation skills, and leading Kaizen events. Project reviews for each student are completed each week by a team of Project Champions. Each student arrives at the class with a chartered project in hand.

- **Week One**
  - **K-Sigma Overview**
    - Card Drop Experiment
    - Roadmap Introduction
  - **Project Definition**
    - Identify Processes
    - Voice of the Customer/Business
    - Identify Metrics and Entitlement—Product Velocity, WIP, Cycle Time, Turns, Throughput, Delay Ratios
    - Identify Constraints
    - Problem Statement
    - Leadership and Financial Approval
    - Project Chartering
  - **Process Mapping**
    - Value Stream Mapping
    - Variables Mapping
    - Swim Lane Mapping
    - SIPOC
    - Handoff
    - Spaghetti Maps
    - Time Value Chart
  - **Current State Assessments**
    - Value/Non-Value Added
    - People, Processes, Products, Materials
  - **C&E—First Level, Steps Only**
  - **Project Management**
    - Change Management
    - Implementation Plan

- **How to Run a "Special K" Mini-Discovery Kaizen Event**
- **Deliverables for Session 2**

- **Week Two**
  - Black Belt Presentation of Session 1 Deliverables
  - C&E Second Level, Variables
  - Intro to Minitab or JMP
  - Basic Stats and Graphs
  - Trend Charts
    - Time Series
    - SPC (I-MR, Xbar-R)
  - Develop Measurement Systems
  - Analyze a Measurement System
    - Data Validity
    - Continuous MSA
  - Attribute MSA
  - Sampling I
  - Systematic, Simple Random, Sub-Group
  - Process Performance
    - Voice of the Customer II
    - Overall Equipment Effectiveness
    - Capability (DPU, DPMO)
    - Capability (Cp, Cpk, Pp, Ppk)
    - Versus Takt
  - Deliverables for Session 3

- **Week Three**
  - Project Charter Update
    - Quantify Entitlement
  - Demand Segmentation, Demand Profile

  - Passive Process Observation
    - Multi-Cycle Analysis and Multi-Vari
      - Analysis Methods (Hypothesis Testing, ANOVA, Chi2, Regression)
  - FMEA
  - Flow Analysis
    - Little's Law, Delay Ratios
  - Application of Previous Tools to Analysis
    - Spaghetti, Time Value, Takt
  - Proper Kaizen—Introduction and Preliminary Preparation

- **Week Four**
  - Presentation: Analysis Results
  - Multi-Vari 2
    - Analysis of Data Set and Multi-Cycle Analysis
  - PRODSim™ Experiment
    - Pull Systems
    - Cellular Design
    - Monument Management
    - SMED
    - Race Car Experiment
    - Product and Process Family
  - Simulation (Flow Charter, iGrafx)
  - DOE (2k, 2k-n,. . .)
  - EVOP
  - Visual Workplace
    - SOP
    - Standard Work
    - Line Balancing
  - Poke-Yoke
  - TPM

  - 5S
  - Kaizen (Proper)
    - Kaizen: Lock and Load
    - Finalize Kaizen Plan
  - Feasibility Study
    - (Cost versus Benefit Analysis of Planned Improvements)
  - Implementation Planning
    - Benefits
    - Owners
    - Action
    - Timing

- **Week Five**
  - Kaizen Event Presentations
  - Design the Implementation Plan
    - Build a Process Control Plan
    - SOP and Standard Work
    - SPC
    - Long Form FMEA
    - Visual Management
    - Dashboards
    - GO Meetings
    - Training/Skills Development
  - Operations Management
    - Roles and Responsibilities
    - Self-Directed Team Creation
    - CI Suggestion Process
    - Balanced Scorecard
  - Audit Planning
    - Control Plan
    - Plan versus Actual Measures/Performances

- SOP
- Cost versus Benefit

- Assessment for the Next Cycle of Improvement
- Next Projects

Six Sigma "Belt" Training Summary

This is a quick overview of the three training programs that are designed to empower action and produce tangible results from Six Sigma. These are all traditional formats, but Six Sigma and Lean can be totally customized and should be customized for each company. For example, we have held Black Belt training for seven straight weeks instead of one week per month for four months. This format immerses the students in their projects and the class posts financial results within 10 weeks instead of the usual 4 to 6 months. This format will work for a relatively small proportion of businesses.

- **Black Belt**
  - Operations (1 week per month over 4 months)
  - Operations (7 consecutive weeks)
  - Product Development (1 week per month over 4–5 months)
  - Services (1 week per month over 4 months)
- **Green Belt**
  - Operations (1 week per month over 2 months)
  - Product Development (1 week per month over 2 months)
  - Services (1 week per month over 2 months)
- **Master Black Belt**
  - Prescriptive: 1- to 2-year development period.
- **Student Evaluations**
  - All events are evaluated on both instructional quality and deployment quality.
  - Feedback is provided within one week of each event.

INTERNALIZATION OF TRAINING

Because of the complex nature of Six Sigma, most companies rely on external consulting groups to kick start the first year or two of the deployments. The goal is to transfer the Six Sigma training skills from the consulting company to the host company as quickly as possible. As soon as a company can say goodbye to the consultants and are able to complete the training with their own resources, the company has internalized Six Sigma. So, training the trainer is always part of a deployment plan. Your Black Belts, with the right development, can train future Green Belts.

The critical milestone on a deployment plan is a Master Black Belt training and development plan. Master Black Belts provide the technical consulting for Six Sigma instead of external consultants. The following is a sketch of a typical Master Black Belt development plan. The first section addresses the role of the Master Black Belts. The second section summarizes the development and training of a Master Black Belt.

THE MASTER BLACK BELT

- **These are the in-house technical experts**, skilled in all facets of Lean Six Sigma:
  - Statistical and Lean tools
  - Project management
  - Change leadership
  - Teamwork
  - Control systems
- **The MBB**
  - **Generates revenue:** Coach and support Black Belts for results.
  - **Develops and delivers** Six Sigma training.
  - **Assists** in project identification.
  - **Partners** with Six Sigma Champions.
  - **Identifies and deploys** best practices.
- **Potential MBBs** rise from the ranks of the BBs who demonstrate strong skills, results orientation, and passion for the tools.

MASTER BLACK BELT TRAINING OVERVIEW

- Master Black Belt candidates are chosen based on
  - Demonstrated ability in achieving results.
  - Interest in driving the Six Sigma program deployment.
  - Certified Black Belt.
- **MBB development program is tailored to needs of candidate and company, and generally contains**
  - Assignment of a mentor.
  - Additional (off-site) training:
    - Statistical methods.
    - Leadership.
    - Other soft skills.
- **Setting of specific goals** for contributing to the success of the program.
- **12–24 month process,** depending on capability and prior experience of candidate.
- MBBs act in this capacity for 1–2 years.

The development of a Master Black Belt is relatively lengthy but when considered as a great leadership development program, the time spent developing Master Black Belts has a good ROI. The development process contains a rigorous certification process and a career path for MBBs. The MBBs should function as a corporate high-performance team that drives the development and delivery of the corporate Six Sigma training plan.

TRAINING PLANS

Training Plan Management. Each division and business in the company should develop training plans. I recommend the businesses include their Six Sigma training plan in the annual operation plan as a requirement. The Executive, Champion, Black Belt, and Green Belt training sessions will be included in this plan. The training courses are customized, linked to strategy, and uniquely designed to ensure that key players are properly equipped to understand and execute their roles and responsibilities. Results won't happen unless effective training happens.

Training plans should be embedded in each person's personal development plan. Training plans should also be developed for continuing education of successful Black

Belts and Green Belts. Continuing education will address additional training in advanced tools and train-the-trainer for training other Belts or peers in the Six Sigma tools. Career advancement plans are developed and directed at Master Black Belt and Black Belt positions.

How Many Do You Train? The question of the number of Belts to be training occurs before the Six Sigma deployment. Figure 11.1 shows the number of Black Belts and Green Belts who were training during the first two years of the Six Sigma deployments for both AlliedSignal and General Electric. Fewer than 2 percent within both companies were trained as Black Belts. Because both AlliedSignal and GE focused on manufacturing operations, the proportion trained as Black Belts may be greater for companies that launch Six Sigma into Operations, Transactions, and Product Development simultaneously.

There was a radical difference in the number of Green Belts that both companies trained in two years. AlliedSignal trained less than 13 percent of their population as Green Belts and GE trained about 27 percent. Why the difference? Jack Welch dictated that every manager should be a Green Belt. That boosted the training numbers but did not radically affect the results.

Both companies brought about $880,000,000 of savings to the bottom line. AlliedSignal easily matched GE's results while training only 3 percent of the number that GE trained. AlliedSignal averaged about $107,000 per trained Belt and GE averaged about $14,000 per trained Belt.

This demonstrated the risk of deploying Six Sigma to a training program rather than focusing on executing high-impact projects with Belt team leaders. I suspect that the executives at GE who were trained as Green Belts didn't complete high-impact projects. But, at the end of the day, I would rather get $880,000,000 using only 8,000 Belts than $875,000 using a huge number of 64,000 trained Belts. Let the process of selecting high-impact projects define the work of Six Sigma Black or Green Belts.

Table 11.2 shows some recommended population percentages based on many other Six Sigma companies. Notice that product design groups have a higher recommendation for people trained. The feeling is that, because product design is the most complex business process technically, more people should be certified Black Belts and every design team member should be a Green Belt.

Table 11.1 Based on Annual Reports, the Number of Black Belts and Green Belts Trained in the First Two Years of the Six Sigma Deployment for AlliedSignal and General Electric

Company	Employees (1,000s)	Black Belts	BBs/100 Employees	Green Belts	GBs/100 Employees	Investment ($ Millions)	Savings ($ Millions)
AlliedSignal	70	1,200	1.7	7,000	10	$33	$880
General Electric	220	4,000	1.8	60,000	27	$575	$875

- Black Belts
  - Operations — 3 to 6 percent of the population
  - Transactional — 1 to 2 percent of the population
  - R&D — 25 percent or higher
- Green Belts
  - Operations — 10 to 15 percent
  - Transactional — 5 to 10 percent
  - Design — 40 percent

Table 11.2 shows a draft of a predeployment training plan covering the first 18 months. It includes leadership, Belt training, and Master Black Belt development. The results reported at the end of the year should match the execution of the training plan. If the training plan is missed, the results will be less.

Table 11.2 An Example of a Six Sigma Training Plan*

Full-Time BB	Part-Time BB
Removed from current position	Remains in current position
Backfill and succession planning is crucial	Current job continues to be done, but at lower productivity
Reports to Champion projects	
Projects can span organization	Reports to old boss
Projects typically in current organization	
100 percent dedicated to Six Sigma projects	50 percent dedicated to Six Sigma projects (actual data shows reality is 5 to 30 percent)

(continues)

Table 11.2 An Example of a Six Sigma Training Plan* (Continued)

Full-Time BB	Part-Time BB
Three to four BB projects per year	One to two BB projects per year
Typically mentor five to ten Green Belts	Rarely mentor Green Belts

*This depicts the first 18 months of the Six Sigma deployment.

Dedicating Time. It's not enough to train a huge number of people like GE did. You must have an infrastructure set to ensure that the trained people do enough to get certified. You also get more ROI if each of your trained people complete several projects over a year. They can't do that part time. Here is my recommendation for dedicating time to your Six Sigma team:

- **Dedicated time to training and project work yields success.**
- **Project time:**
  - The more time dedicated to a project yields higher returns in a shorter period of time.
  - NOTE: This has been shown in benchmark data from corporations already involved in similar programs.
- **Benchmark companies:**
  - Champions: half hour per week per BB
  - Master Black Belts: 100 percent
  - Black Belts: 70 to 100 percent time
  - Green Belts: 20 to 50 percent of their time

Simply, the more projects completed, the better the results. Also, the more time dedicated to project work, the more projects are completed. If your project selection process identifies great projects, you would be shooting yourself in the foot by having a bunch of part-time Belts running around. A full-time Black Belt should be able to complete four or more projects per year. Each Black Belt should be worth $1,000,000 of savings per year if deployed aggressively. This is a great way to develop your future leaders.

Student Evaluations. The final section on training addresses student evaluations. Student evaluations are as much a part of the Six Sigma infrastructure as anything else. The Six Sigma Steering Team requires evidence that the quality of training is high and consistent. Feedback from the evaluations leads to courseware modification that fits the company.

The student evaluations provide performance data on your instructors and also provide great information about the progress of your deployment. The student evaluation has two components: (1) the quality of the training and (2) the quality of the deployment. The following is an example of a student evaluation.

You will see there are three sections to this evaluation. The first section assesses the instructional quality of the course itself (Items 1–20). The second section allows the students to rate their perception of the usefulness of the tools taught to them. And, the third section addresses how things are going for each student from the deployment standpoint. Is their Champion supporting their efforts? What percentage of their time are they spending on their project? And, are they getting frequent project reviews?

BLACK BELT COURSE EVALUATION—WEEK 4

Course Name: **Date:**

Instructors' Names:

Please circle the number that best expresses how you rate each of the evaluation criteria listed.

Note: 5 is the highest score; I is the lowest.

Instructional Quality	**LOW**				**HIGH**
1. Objectives for the course were clear.	1	2	3	4	5
2. Course objectives were met.	1	2	3	4	5
3. Course material was of practical value for my job.	1	2	3	4	5
4. Course material was generally presented clearly and accurately.					
Instructor One	1	2	3	4	5
Instructor Two	1	2	3	4	5
5. Instructor was enthusiastic about the subject.					
Instructor One	1	2	3	4	5
Instructor Two	1	2	3	4	5

6. Instructor was knowledgeable about the subject.					
Instructor One	1	2	3	4	5
Instructor Two	1	2	3	4	5
7. Instructor allowed me to ask questions and be involved in the discussions.					
Instructor One	1	2	3	4	5
Instructor Two	1	2	3	4	5
8. Instructors gave me adequate time to provide course feedback.	1	2	3	4	5
9. The course sequence was logical.	1	2	3	4	5
10. Group exercises were meaningful.	1	2	3	4	5
11. Adequate time was provided for group exercises.	1	2	3	4	5
12. Amount of time allotted for class was appropriate.	1	2	3	4	5
13. Facilities were adequate for the training session.	1	2	3	4	5
14. The degree of confidence I have that I'll use the skills and knowledge gained in this course.	1	2	3	4	5
15. I would recommend this course to others.	1	2	3	4	5
16. My overall assessment of this course.	1	2	3	4	5
Instructional Quality for All Four Weeks	**LOW**				**HIGH**
17. Week 1: Measure (Process Maps, C&E Matrix,. . .)	1	2	3	4	5
18. Week 2: Analyze (Stats, Multi-Vari, ANOVA,. . .)	1	2	3	4	5
19. Week 3: Improve (DOE,. . .)	1	2	3	4	5
20. Week 4: Control (RSM, EVOP, Control Plans,. . .)	1	2	3	4	5

Please Rate the Usefulness of Six Sigma Tools	LOW				HIGH
1. Process Map	1	2	3	4	5
2. C&E Matrix	1	2	3	4	5
3. Measurement Systems Analysis	1	2	3	4	5
4. Capability Studies	1	2	3	4	5
5. Failure Modes and Effects Analysis	1	2	3	4	5
6. Multi-Vari Studies	1	2	3	4	5
7. Fractional Factorials	1	2	3	4	5
8. Full Factorials	1	2	3	4	5
9. Response Surface Methodology	1	2	3	4	5
10. Evolutionary Operations	1	2	3	4	5
11. Control Plans	1	2	3	4	5
12. Statistical Process Control	1	2	3	4	5

Implementation Quality	LOW				HIGH
1. There is clear leadership support for your project.	1	2	3	4	5
2. Your team gets the right support from other operational or functional groups.	1	2	3	4	5
3. My Champion provides the right level of support.	1	2	3	4	5
4. The probability of success of my project is:	1	2	3	4	5

5. What percent of your time have you spent on your project? __________Percent

6. What percent of your project is completed? __________Percent

7. How many times has your project been formally reviewed since the last week of class? _________Times

8. What is the updated forecast of the financial value of your project? $_________

9. How much has your project saved to date? $__._______

10. What two things did you like best about this course or program?

____

____

11. What did you like least about this course or program?

____

____

12. Please make any additional comments or recommendations in the remaining space or on the back of this form.

____

12 Communicating the Six Sigma Program Expectations and Metrics

With Joe Ficalora, Roger Hinkley, and Joyce Friel

You have now successfully planned your Six Sigma program, and your series of leadership workshops are complete. The next challenge is to plan the effective communications to your organization about the new initiative. In this section, we will explore what works, and how to say it. The strategy should be a series of simple messages repeated over and over and delivered throughout the organization by a wide variety of media. The future belongs to leaders who can create and then communicate vision. At best, communicating the Six Sigma initiative will be a team effort including every leader involved with Six Sigma.

Creating a compelling message illustrating exactly why Six Sigma is important to the future of the company is not an easy task, but provides the first communication challenge. A good communications plan will also consider a wide variety of communication methods. One Fortune 500 company that deployed Six Sigma used the following:

- CEO and other Executive Team Members were consistently the keynote speakers at any Six Sigma event
- Global broadcasts
- Quarterly satellite broadcasts
- Videos and MPEG messages
- Coaching Black Belts to communicate effectively

- Coaching Black Belts to create a mystery story to describe their project
- Brightly colored posters

This chapter will cover a wide variety of considerations in the communication of a change initiative. To say this is an art may be an overstatement. Leadership communication is a fine science in which there is little room for error. Two simple rules are: Create a simple message, and communicate it repeatedly. Larry Bossidy once said he knew when he communicated a change enough when he got to the point of feeling like he would throw up if he said the message one more time. This chapter will cover the following:

- Creating the message
- Selecting communication media
- Developing a communication plan
- Customizing the message
- Evaluating the communication effort

A COMMUNICATIONS MODEL

Joyce Friel, President of Peak Performance Consulting, has provided a simple model for evaluating and delivering messages. This model results in a 2 × 2 matrix. The two axes of the model are message complexity and message emotional content. Create a message and decide in which of the four resulting quadrants the message fits. Any given message may be simple or complex to understand, but the degree of emotional reaction/content to any given audience might be completely different. So the same message might be delivered differently to the two or more different groups.

Low Complexity and Low Emotional Content. Messages in this category can be delivered in any number of formats (mass media, public forums, verbally, electronic, or written) because it is easy to understand and the content is not emotional. For example, the company has declared the day after Thanksgiving to also be a company holiday, or the operations team in a particular sector will be using the West break room for their Six Sigma milestone celebration on Thursday between 3:00 and 4:00 p.m.

High Complexity and Low Emotional Content. Messages in this category need to be in written and verbal form because the content is complex, but doesn't carry any emotional content. These messages should also be delivered multiple times because of the complexity. For example, the calculation of a particular metric (Rolled Throughput

Yield) would fit in this category. For those not comfortable with statistics and formulae, this message could be confusing and complex, but for engineers, the message might not be so complex. Motorola, for example, constantly communicated to employees the basics of calculating Sigma many times over a five-year period.

Low Complexity and High Emotional Content. These messages need to be delivered verbally several times because the more emotional the content, the less of the message we actually hear the first time. We tend to only hear the part that was spoken before our emotions kicked in. So you have to say it over and over again and in different ways until the message gets through the emotion. Also, it is wise to deliver these categories of messages in small groups to appropriately accommodate the emotion that may be displayed.

Be prepared for this show of emotion. Messages such as learning that your job is being changed to support a breakthrough improvement initiation like Six Sigma will tend to be highly emotional. The change in the job may be very simple and even beneficial both to the company and the employee, but because employees are so vested in their jobs and feel so threatened when they change, the emotional content may overshadow the simple change in the job structure or content.

High Complexity and High Emotional Content. These messages are the most difficult to communicate, being weighed so heavily in both complexity and emotion. Such messages have to be delivered many times in many different ways in small groups and one-on-one. For example, the company has decided to launch Six Sigma; they are starting with your division, and you have been selected to be the Deployment Champion. This message is complex because you don't understand it yet, and the message is emotional in that the best leaders are usually picked, and those who are picked are viewed by both themselves and their peers as being the best leaders. The chosen don't want to risk failing, but feel they might fail by taking on such an unknown role.

These are the kinds of emotions that are often not spoken, but always felt. Many of the messages you deliver during the first 90 days will fall in this category of high complexity and high emotionality. This category impacts your communications plan because these messages must be communicated much differently than messages in the other three categories.

Table 12.1 summaries this model with communications tips for each quadrant.

Table 12.1 Communications Matrix, a Guide for Deciding the Type and Frequency of Communication

High Complexity	Written Format Verbal Format Opportunities for Questions Multiple Time	Written and Verbal Video if Necessary 1:1 or Small Groups Anticipate and Prepare for Emotion Multiple Times and Formats
Low Complexity	Mass Media Either Written or Verbal	Verbal Generally Tell Several Times 1:1 or Small Groups Allow and Prepare for Emotion
	Low Emotion	**High Emotion**

CREATING SPECIAL COMMUNICATION SYSTEMS

Companies customarily use their usual communication avenues to communicate the impending initiative, but sometimes it may be suitable to develop new communication programs to augment the more established ways. For example, in 1995, the engineered material sector of AlliedSignal started a program called "Winning Together." Winning Together was a simple program that proved to be quite effective. The program had three objects, as follows:

1. Establish the Six Sigma metrics throughout the sector.
2. Educate and communicate the Six Sigma program and tools.
3. Recognize success.

Soon after launching the Winning Together program, each factory had a Winning Together board installed at the reception area of the site. This board showed the status of the site's program on several metrics:

- **Six Sigma Metrics**
  - Rolled Throughput Yield (RTY)
  - Cost of Poor Quality (COPQ)
  - Capacity Productivity (C-P)

    - Customer Satisfaction (On-time delivery)
    - Safety (OSHA metrics)
- **Business Metrics**
    - Sales
    - Income

and presented the status of each metric using red and green traffic lights. A green light displayed if you were better than goal, and a red light displayed if you were not. If a site got a threshold amount of green lights, then everyone at the site was eligible for special prizes and gifts. Figure 12.1 shows how the Winning Together board was linked to the goals of Six Sigma.

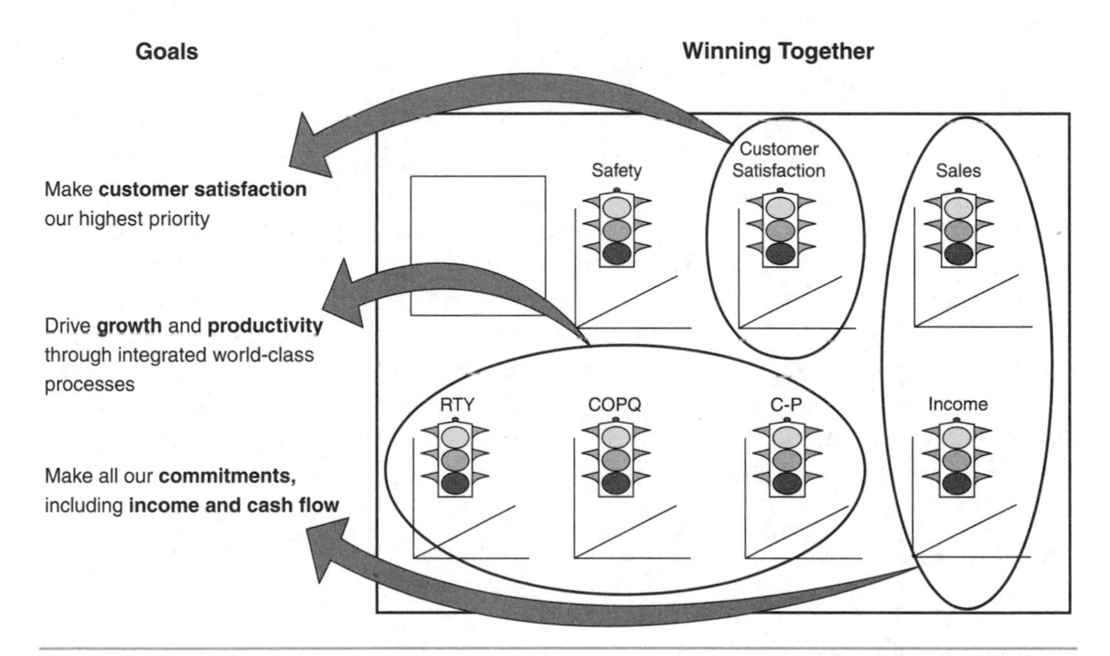

Figure 12.1 The Winning Together board that shows the status on several metrics for a manufacturing plant.

The Winning Together program had the purposes of communicating Six Sigma goals (scoreboards), getting everyone involved (monthly meetings), and recognizing success (quarterly recognition). This was a process and system developed specifically to tie Six Sigma to everyday life on site. Figure 12.2 shows a status board used in the sector operating reviews.

Simple Summary of Q3 Metrics

	Division A	Division B	Division C	Division D	Division E	Company Summary
RTY	Red ●	Red ●	Green ●	Red ●	Red ●	Red ●
COPQ	Green ●	Green ●	Green ●	Green ●	Green ●	Green ●
C-P	Green ●	Red ●	Green ●	Green ●	Green ●	Green ●
YTD Savings	$23,723K	$14,188K	$4,626K	$4,690K	$5,075K	$52,302K

Figure 12.2 A status board for the divisions of the engineered materials sector of AlliedSignal.

The Winning Together program was specific to Six Sigma, but would have worked with any initiative. The program was implemented quickly, and at least one plant manager lost his job when he refused to put up a Winning Together board in his plant. The plants took the program a step further. In the cafeteria of many of the sites, all the Six Sigma projects and project teams were put on the cafeteria wall with the goals that the projects impacted. This was an excellent way to get visibility on the project work.

The engineered materials sector of AlliedSignal used a wide variety of communication methods as well. After each class of Black Belts was completed, color posters were developed to include a project summary and a picture of the Black Belt who lead the project. These posters were shown at the sector headquarters in Morristown, NJ, and then sent to the Black Belt. Communication kits were sent out periodically and large leadership conferences focused on Six Sigma were held frequently.

The Message

I referred to John Kotter's stages for change in Chapter 3, "Six Sigma Launch Philosophy." Of the eight stages, the fourth stage is titled, "Communicating the Change Vision." Kotter defines seven principles in communicating the change vision:

1. **Simplicity**—Low word count; no jargon.
2. **Metaphor, analogy, and example**—Create a verbal picture or a physical picture.
   a. Fred Poses, when President of the engineered material sector at AlliedSignal, created the sports car, the Viper, as the symbol of Six Sigma. Driving that car up the

hill to world class results. He gave a few employees a chance to drive a Viper pending dramatic results.

3. **Multiple forums**—Use every forum available. Create new forums if necessary.
4. **Repetition**—Classic communication technique. Even Shakespeare used it in his drama.
5. **Leadership by example**—Emotional investment really counts.
6. **Explanation of seeming inconsistencies**—"We did TQM, so why Six Sigma?" One company used the motto, "Six Sigma puts the brains into TQM."
7. **Give and Take**—Two-way communication is always good. Elevator speeches, short speeches that capture the essence of Six Sigma, are great.

One of the best messages I ever heard was from the president of a medium-sized chemical company at the executive kickoff. He listed three current metrics, tracked by the organization enterprise-wide, where they stood and where they were going to be by year's end. They were as follows:

- Warranty returns: from 1.5 to 0.2 percent
- Overall process yields: from 82 to 88 percent
- On-time delivery: from 97.5 to >99 percent

He further went on to say that by doing all of these, the organization would grow in customer orders and have higher profits, and the previous three-year trend of layoffs could be stopped or reversed. He really had everyone's attention at that point. Each of his main goals was grounded in data, backed up with customer surveys and employee surveys. He had done the hard work of preparing in great detail to deliver a simple message.

- **The Message**
  - **Who**
    - Each function and level in the organization gets the same message.
    - The leaders of each function should absolutely be required to deliver the message in person to their charges.
    - Timing in terms of organizational hierarchy may be required, but do not let the message get distorted by word of mouth.
    - A written, bulleted message will anchor the gist of the message and prevent distortion as it passes from person to person and from level to level within your organization.

- **What**
  - The Six-Sigma rollout message must state specifically what is expected overall and then in detail what will happen.
  - This is the part of the message that must be succinctly captured at the beginning and then explained in reasonable detail so each part of the organization can see the part they will play.
- **When**
  - The timing of the message needs to be considered from several viewpoints.
  - Who needs to know about this and when do they need to hear it?
  - Obviously the organizational leadership needs to know, but they may want to be part of drafting the message.
  - In the latter case, you may elect to make that part of an executive kickoff session deliverable.
  - In the former case, think through what your external reality is telling you and how you wish to confront it. Then get that message out to your top leaders right away.
- **Where**
  - Where should this message be delivered? Consider that the most effective place to deliver it is at the workplace and shop floor.
  - Company-wide meetings are appropriate venues, especially if an annual or quarterly meeting is scheduled in the near term.
  - If you want extraordinary results from this effort, unplanned visits in person will let people know just how extraordinary this effort will be!
- **How and How Much**
  - Everyone will want to know exactly how this rollout is planned.
  - How much commitment do you personally have for this initiative?
  - People will commit only as much as they see their leader committing to Six Sigma. They will be wondering exactly how much effort, resources, and time you are expecting from them.
  - Be as precise as you can be here—how many Belts are to be trained?
    a. How much time will they be spending on projects?
    b. Full time or part time?
    c. How much savings are expected from each project?

   d. Each Belt?
   e. How soon will the program get started?

- **Burning Platform or Burning Desire?**
  - Every leader knows his or her organization's culture. What is your culture?
    a. Do people seem happy and motivated when it comes to new initiatives or changes?
    b. Or are there past instances when things have not gone so well with change efforts?
  - As a leader, you must assess whether your organization's external reality is one where change must be driven immediately or whether you are doing this to improve faster than your nearest competitor.
  - If it is a "burning platform," then that must be part of the message delivered.
  - If it is a desire to continue a string of successful changes within the company, then that must be highlighted.
  - Either way, the message must be stated in a positive way, with a good vision of where you are leading the organization. A powerful vision, with a positive message about how to get there, is a great motivator.
- **Market Challenge and Vision for the Future**
  - Current and perceived future market challenges be included in the message.
  - The Ford Automotive Company, many years ago, had a message that showed how once an automobile company slipped to fourth or fifth place in the U.S. market, it was on the path to extinction. Coupled with such a message was no doubt a path to changing the company.
  - Ford has changed dramatically in the past five years. Their trucks and SUVs own a significant market share, in spite of European and Japanese challenges. Their factories are both flexible and efficient.
  - Their designs are driven by the use of common subassemblies, architectures, and platform planning. These approaches help make Ford a formidable competitor in their marketplace, rising to market challenges.

The following is an example of the message sent by the Executive Team to the senior leadership of the company. It is simple and exciting and certainly gets the deployment requirement out front.

SIX SIGMA PROGRAM DEPLOYMENT

- To achieve our goals, we must:
  - Believe—We have a great opportunity.
  - Find—Be thankful but not embarrassed.
  - Communicate—Make it visible.
  - Deploy—Put it in the hands of our people.
  - Execute—Focus, focus, focus.
  - Track—Use the new measures.
  - Reward—Recognize higher performance.

Each leader must communicate the company's market challenges and their personal vision for the future. The vision must be communicated in simple language that translates easily across all organizational levels, to everyone in the company. The vision must be positive, as people are drawn to something positive that is reality based. Fantasy is for the movies, and the moment anyone believes a vision is more fiction than fact based, you will have lost them.

THE MEDIA

Paint the future state in as much detail as possible, and you should be able to clearly describe the steps to get there. Kouzes and Posner postulate that leaders live their lives in reverse. They know exactly where the organization will be in the future, and then walk backwards in time to develop the steps necessary to get there. So make certain your vision is positive, detailed, and with enough motivational facts to get people moving. Finally, deliver it with high energy in your personal style to show your commitment. Now let's take a look at some media for communications.

- **The Media**
  - **Intranet**

1. You may want to post your initial communication of Six Sigma and the planned rollout on your company intranet, if you have one.
2. Update it monthly or at least quarterly with progress to plan comments.
3. This is a great place to showcase those early successes to continue developing a vision of how it is changing the company and exactly what progress looks like.

4. Some people need to see concrete results or case examples before they can understand fully what value Six Sigma can bring to your organization.

- **Publications**
  1. If your company does regular newsletters to employees and stockholders, this is a great forum to showcase some key messages, both initial and progress updates.
- **The Chairman's or President's Message**
  1. If you regularly communicate in these forums, Six Sigma must be a significant part of it.
  2. It cannot be a casual mention, and it should be listed first above all else if you want people to sit up and take notice.
  3. Bob Galvin of Motorola would sit down in company reviews and leave after the Six Sigma and Quality portions of the reviews were completed. People got the message of priority.
- **Video**
  1. A professionally done video done by the CEO or senior executive is effective.
  2. Jack Welch was known for his video performance at the beginning of GE's Six Sigma deployment.
- **Briefing Meetings**
  1. Briefing meetings lead by the site leadership are effective for their face-to-face venue.
  2. Meetings should be carefully scripted with professionally developed briefing materials.
  3. Audience and frequency of meetings should be carefully considered.
- **Recognition Events**
  1. Companies have had special events to honor participants.
  2. Poster events—teams bring project posters and are available for questions and answers.
  3. AlliedSignal Fifth Session—all Black Belts in training were invited to a celebratory session one month after their training was completed.

COMMUNICATION PLANS

Early in the deployment of Six Sigma, a communication plan is developed to cover the first six months of the launch. A plan for a single site might be fairly simple. Here is an example of one:

Six Sigma Communication Plan; Local Organization

[**Week 0**] Internal communication (newsletter).
[**Week 1**] Meeting review (every week); metrics/goals definition for the plant/area.

- Project definition (who, when, how).
- Review project plan and metrics (due dates).
- Internal communication (newsletter).
- Initiate safety investigation group.
- External communication between plants/areas.
- Start safety communication boards (plant/area entrance).
- Each team can use other persons to meet the goals.

[**Week 2**] Internal communication (newsletter and publish project schedule on site).

- Hands-on project implementation.
- Review metrics accuracy.
- Prioritization, accountability.
- Safety investigation group project.
- Meeting review (every week).

[**Week 3**] Internal communication (newsletter and publish metrics results on site).

- External communication.
- Meeting review (every week).

[**Week 4**] Internal communication (newsletter and publish metrics results on site).

- External communication (boards of directors).
- External communication between plants.
- Safety investigation group results.

- Recognition to the new Six Sigma personnel.
- Meeting review (every week).

A team drafts the communication, and the team includes at minimum the PR folks and the HR folks. I tend to think of the questions I want to answer before launching a communications plan, such as the following:

1. What results do I want from the communication?
2. How will I know that the communication was successful?
   - Metrics (defined before the communication goes out)
   - Method for capturing and analyzing the metric data
3. What do I need to convey in the communication to achieve these results?
   - What information will the team need to know to achieve success?
     - i. Strategy
     - ii. Business plan
     - iii. Timing
     - iv. Value network
     - v. The framework for:
       1. Identifying and responding to customer needs
       2. Solving problems
       3. Gathering input
       4. Reacting to competition
       5. Striving for strategic and business goals (revenue, profit, cash flow, etc.)
   - Roles and responsibilities
   - Reward and recognition plan
4. When putting together the communication package, consider an advertising campaign:

   Herb Krugman of GE suggests that a minimum of three exposures per customer is required to impact the customer (they either act on it, get irritated with it, or forget it). The first exposure is unique. A "What is it?" cognitive response dominates the reaction. The second exposure produces several effects: "I'm interested" or "I'm not interested" are among them. The third exposure constitutes a reminder. The draft process for doing this is as follows:

- Introduce key concepts on launch and promotional planning.
- Review targeting and positioning considerations.
- Discuss management and statistical tools available to prepare for launch activities.
- Just as with preparing sales teams for a new product or campaign, your team/employees
  - Are likely to be the greatest skeptics.
  - Already have
    - Their own ideas.
    - Their own experiences.
    - Their own biases on what will and will not work.
  - Will typically do what they believe is best to maximize their concept of success.

1. **Positioning—Create a unique place for the program in the employee's mind.**
   - **Message**—What we need to say to communicate positioning.
     - Use words to tell the story.
     - Includes data when appropriate.
     - Don't confuse things by trying to look at graphic images at the same time you are developing "voices" for the message.
     - Develop a "best guess," and then refine the message to gain the team's mind (buy-in).
     - Focus on what the team needs to hear/know.
   1. Decide which "elements" to include.
      - Start with the positioning concept.
      - Add important elements from the product profiling efforts.
   2. Decide the emphasis for each element.
      - Emphasis (weight) affects how well it will be retained.
      - Decrease or increase emphasis by adding or subtracting specific details, data, and so on.
      - Use team/customer scoring methods to decide the emphasis on the already determined elements.
   3. Decide the order of the elements.
      - Don't always follow traditional wisdom.

  - Get your main point(s) of difference across first, then follow with other important elements.
4. Decide the use of supporting data for the elements.
  - Use data judiciously in promotion.
    - Avoid information overload.
  - Use detailed, reliable, credible data only to support KEY message elements.
    - Reserve other data for back-up/follow-up materials.

2. **Communication has been done right when the employees/teams**
   - Demonstrate clear understanding of the communication
   - Can provide overall comments about the communication and surrounding information.
   - Can identify and relate to each and every element.
     - Clear comprehension of positioning.
     - Relevance to processes, programs, and deliverables.
     - Believable program differentiation.
     - Appropriateness of documentation presented.
     - High interest in program.
     - Understanding of how/when to implement the program.

Finally, even the Black Belts and Green Belts need to know how to communicate. Here is an example of Black Belt project drivers. These are the communication suggestions that the Black Belts will receive during their Black Belt training.

Black Belt project drivers:

- Recognize team often and publicly.
- Put team picture on the wall.
- Keep them apprised of outcome of trials.
- Schedule team for presentations to plant leadership.
- Keep their current activities on the wall in area.
- Include the team in celebration of success.
- Give tangible item of membership (hat, jacket).
- Send thank-you letter home to share with family.

Elevator Speeches. An effective way to communicate a change initiative is the elevator speech. When we presented sample agendas for leadership workshops in Chapter 10, "Creating Six Sigma Executive and Leadership Workshops," there was usually a section set aside to develop the elevator speech. This speech is short and simple. If you get on an elevator at the top floor of the corporate headquarters, and the CEO gets on the elevator with you and asks, "Tell me what you think of Six Sigma," you have until the elevator gets to the ground floor to make your point. Consistent elevator speeches enhance the chances the ideas will be repeated a lot and consistently.

When working at AlliedSignal, I always had two elevator speeches ready: (1) the status on the Six Sigma deployment and (2) recent financial results. Here are some actual examples of some Six Sigma elevator speeches.

Six Sigma Example Elevator Speech #1

- **The Right Metrics**
  - Customer satisfaction
  - Defects per unit or rolled throughput yield
  - Cost of poor quality
  - Cycle time/rate
- **The Right Projects**
  - Close the gap between baseline and process entitlement
  - Aggressive project review structure
- **The Right People**
  - Champions—catalyst for change
  - Master Black Belts—internal technical experts
  - Black Belts, Green Belts
- **The Right Roadmap and Tools**
  - Measure, Analyze, Improve, and Control
- **The Right Results**
  - $$ to the bottom line—measured by financial group

Six Sigma Example Elevator Speech #2

- **Why**—Reduce variability in all processes to achieve the AOP and SBP while providing breakthrough business results.

- **Who**
    - Champions, Master Black Belts, Black Belts, and Green Belts.
- **How**
    - Plan, Train, Apply, Review.
    - Define, Measure, Analyze, Improve, Control.
- **Where**—Every business; every function.
- **When**—The train has already left the station.

Six Sigma Example Elevator Speech #3

- Fosters "one-company" vision.
- Provides common approach, language, and methodology across all companies and functions:
    - Step-by-step common language roadmap.
    - Design, manufacturing, and services common tools.
- Ensures focus on breakthrough performance.
- Infrastructure:
    - Reviews at Champion, Plant, and Executive levels.
    - Champion and Master Black Belt network.
- Project-specific action learning with review.

Six Sigma Example Elevator Speech #4
Our Six Sigma Commitment to Drive

- **Growth**
    - Understanding the voice of the customer
    - Value proposition
    - Faster technology development
    - Faster product commercialization
- **Cost/Productivity**
    - Quality improvements
    - Cost of poor quality
    - Capacity improvement (without capital)

- **Cash/Working Capital**
  - Payables
  - Receivables
  - Inventory

The communication of the change initiative is a challenge. But your company has probably done this before with different initiatives. Some of them were well communicated and some not. Replay those successfully communicated initiatives and use those initiatives as your planning starting point. Topics for your initial communications plans over the first three or four months might include the following:

- Six Sigma linked to strategy
- Roles and responsibilities:
  - Initiative Champion
  - Deployment Champion
  - Project Champion
  - Black Belt
  - Green Belt
  - Master Black Belt
- Training programs
- Training plan
- Six Sigma methodology
- Projects:
  - Selection
  - Prioritization
  - Tracking
- Six Sigma related to previous initiatives

Communicating Six Sigma has a learning curve built in. The leadership of the company must understand the basics of Six Sigma, and training events early in the deployment are designed to do that. Remember: Be simple, be creative, be visual, and be repetitive. Holding to those principles will get you there.

PART III
POST-LAUNCH

13 Creating the Human Resources Alignment

With Kristine Nissen and Joyce Friel

The role of Human Resources (HR) is varied across companies. But, when launching a change initiative, the HR arm of the company can take a leadership role in that initiative. When considering a Six Sigma deployment, several parts of the deployment plan are directly linked to HR. Although HR leaders are not usually billed as deployment leaders, the HR leaders can still drive the quality of the deployment by supporting the new Six Sigma infrastructure to do the right things.

A Six Sigma deployment includes defining and implementing a supporting infrastructure, including steering teams, Initiative Champion, Deployment Champions, and Project Champions. In addition, the company creates almost overnight literally hundreds of Master Black Belts, Black Belts, and Green Belts. Now creating and leading steering teams and project teams becomes the priority.

So, HR must look change right in the eye and not back off. Six Sigma deployments have required the realignment of almost every organization in the company and the creation of completely new job positions and expectations along with new career ladders. HR can readily impact the company's ability to lead teams and develop new coaches and mentors (Master Black Belts and Black Belts) throughout the company. Even the rewards and recognition systems will be modified to reflect the new culture change.

And finally, HR departments can also lead the company during the early days of Six Sigma deployments in driving Six Sigma process improvement on their own processes. Here is a rough list of no fewer than 21 different processes HR commonly manages:

1. Employee benefits/payroll/HR
2. Hiring and selection process
3. Orientation
4. Annual enrollment—benefits
5. Relocation
6. Terminating/transfer employees/severance
7. Plant closure and acquisition
8. Performance reviews
9. Salary increase/budget
10. Paying employees/retirees
11. Employee incentive programs
12. Retirement process
13. Education and training of employees
14. Reward and recognition
15. Recognition service awards
16. Union negotiations
17. Managing expatriates
18. Record keeping
19. Government compliance and reporting
20. Vendor management
21. Results through people

To have all these processes simultaneously operating at world-class levels definitely takes a few Black Belts and Green Belts doing a lot of projects. What better way for HR to take a leadership role in the deployment than taking a leadership role in applying Six Sigma to their processes?

HR is complex and multifunctional. There are many ways HR impacts a new initiative's deployment quality. From a Six Sigma deployment viewpoint, some of HR's general roles are as follows:

1. Assist in personnel selection.
2. Develop workforce practices that enable Black Belts and Green Belts to achieve high performance in their workplace.
3. Work with executives and managers to establish appropriate recognition and rewards.

4. Develop Six Sigma education and training support that will contribute to employee performance.
5. Assist in establishing appropriate career plans and career paths for Master Black Belts, Black Belts, and Green Belts.
6. Track key players (Champions, Black Belts, and Green Belts) for "success profiles."
7. Provide input to the personnel selection and future hiring process.
8. Work with Business Leadership to ensure
   a. Consistently successful Black Belts and Green Belts are recognized and developed further.
9. Assist MBBs, Black Belts, and Green Belts in developing their training plans.
10. Assess organizational readiness for change.
11. Sponsorship assessment.

In this chapter, I will address several opportunities for HR to take a clear leadership role in Six Sigma change. As you can see, the HR role in making the Six Sigma deployment smooth is critical.

- Talent selections
- Job descriptions
- Training plan development
- Succession planning
- Reward and recognition
- Career planning
- Compensation
- Retention and career planning
- Deployment support

SIX SIGMA ORGANIZATIONAL STRUCTURE

The roles within the Six Sigma deployment infrastructure include the following:

- Steering Committee
- Champions

- Process Owners (Project Sponsors)
- Black Belts
- Master Black Belts
- Green Belts
- Team Members

The roles that are most amenable to HR support are those of Master Black Belts, Black Belts, and Green Belts. These are essentially new positions within the company. Their roles have been discussed in Chapter 8, "Defining the Six Sigma Infrastructure," so be sure to review those. A quick review of the Belt roles follows.

Master Black Belts

- Technical leader in Six Sigma, methods, and tools.
- Respected at all levels.
- Coaches and supports Black Belts for results.
- Delivers Six Sigma training.
- Assists in project identification.
- Partners with Six Sigma Champions.
- Identifies and deploys best practices.
- Participates in a one-year development program.

Black Belts

- Change agents for institutionalizing Six Sigma.
- Respected across the organization.
- Proven hard and soft skills.
- Leads strategic, high-impact process improvement projects.
- Masters basic and advanced quality tools and statistics.
- Deploys techniques of measurement, analysis, improvement, and control.
- Participates in intensive four-week training program.

GREEN BELTS

- Change agents for institutionalizing Six Sigma.
- Respected within their organization.
- Proven hard and soft skills.
- Participates in strategic, high-impact process improvement projects.
- Masters basic quality tools and statistics.
- Helps deploy techniques of measurement, analysis, improvement, and control.
- Participates in a two-week training program.

Several issues reside with the creation of Belt positions and the selection of Belt candidates. These issues include whether to develop Belts internally or hire them from outside the company. Once that issue is resolved, the next issue is: Are the Belts full time on project work or part time—this topic is usually very controversial. And, finally, the last question is: How many people are to be trained over the first year? Rough forecasts of the total training volume are necessary to put together a good annual training plan.

TALENT SELECTIONS

Develop Black Belts and Master Black Belts or Hire from the Outside? There is a tendency in some companies launching Six Sigma to try to kick-start the program by hiring a bunch of Black Belts and Master Black Belts from other companies. In a sense, they try to buy the change program. The action might seem to make sense because these new resources are ready to go. But you can't ignore the reality that Black Belts can be trained and launched in four months or less and Master Black Belts in a little over a year. I worry about the message you are sending if you bring in a lot of new people from different companies. Current employees may be thinking, "Why wouldn't the company invest in us, and who are these people?"

Because Six Sigma is customized for each company, the training curricula holds commonality with the training in other companies, but the training will not be identical. So, even though you hire a trained Black Belt from another company, you will invariably have to retrain him or her. The new resources will have to understand the new roadmaps and tool sets that your company is using. They will, unfortunately, have a tendency to second-guess any difference in your approach because their previous approach worked for them and must be better.

From the consulting side, I have seen trouble occur from these newly hired resources even to the point of sending the new resources on to other companies. My strong feeling leads me to believe it is much better to develop your own people first. If you find gaps in capability, then it may be time to hire from the outside. There exists a range of skills in Six Sigma that are found across the industry. Subtle and major differences in the training curricula exist, so a Black Belt from Company A is not necessarily the same as a Black Belt from Company B.

For example, the original Master Black Belts within GE went through a two-week training program, and that was it—poof, they were MBBs. Alternately, the Master Black Belts initially developed within AlliedSignal accomplished a one-year development plan that included weeks of training in advanced statistical tools, plus training boot camps to hone their training skills. There was a radical difference in the capabilities of the GE Master Black Belts compared to the AlliedSignal Master Black Belts. The GE Master Blacks were more involved in Six Sigma deployment activities than mentoring Black Belt and Green Belt projects.

I have seen other differences in Six Sigma training even within a large company. One division might conclude that three days is all that is needed to train a Green Belt, and another division might stick with the traditional two weeks of training. The lesson learned here begs you to perform a thorough assessment of the skills for any Belt hired in from another company. This is where your Six Sigma consulting company may be able to help. The consultants can interview candidates and give you objective feedback concerning the match of skills to your program.

External hires will always have different understandings of the Six Sigma program and have different paradigms for improvement. To add to those inconsistencies, it takes longer to hire from outside and ramp them up when compared to growing your own BBs. Add to that the more important dynamic—your own employees feel slighted, and not as appreciated or invested in. I recommend you grow this program internally. The HR function will have a significant influence on the direction the Six Sigma Steering Team takes on this issue. It is best to think about it ahead of time.

Do the Belts Work Full Time on Project Work or Part Time? The first impression that leadership gets in the early days of a Six Sigma deployment is that with all these Belts running around doing projects, how are we going to hire enough extra people to get the job done? Whether to have BBs work full time on project work is always a serious question early on. Table 13.1 shows a comparison of full-time and part-time Black Belts.

The tendency is to think full-time BBs must be backfilled. I suspect that full-time BBs can remain in their current position, but parts of their former responsibilities must be delegated. Full-timers might report to the Champion and part timers would report to their old boss. So there might be a difference in the new reporting structure. The big difference between full- and part-time BBs is that the full-time BBs are 100 percent on their projects and the part-time BBs are less than 50 percent time on the projects. My data shows the percent of time on project for part-time BBs is anywhere from 5 to 30 percent.

Because of the time allocation, full-time BBs are expected to complete three to four projects per year, and part-time BBs will be lucky to get two projects done. So, what's important? Maintaining the old culture and miss opportunities to improve, or commit to the new process and drive your business strategically in the right direction?

My sense is that all the controversy over full time versus part time is a waste of time. If your leadership group commits to doing a great job in selecting high-impact projects, the nature of the projects themselves will drive the decision about resource allocation. If you've identified a project that your financial people say will bring over $1,000,000 to the bottom line, you'd be shooting yourself in the foot by allowing the BB to work only 30 percent of his time on that project. If you pick great projects the right way, allocation takes care of itself. From the HR standpoint, you will help your leadership clarify the model that best benefits the company and the individual.

From my AlliedSignal experience, we did a great job of selecting projects and assigning resources. Almost all the BBs in my program were part time, but worked their important projects full time. Select good projects and make sure they get done! All good leaders do that.

To help the organizations participating in Six Sigma answer the question about full versus part time, have the Six Sigma stakeholders perform these exercises as a group. This exercise should take about 45 to 60 minutes.

- **If your company has not yet decided whether to use full-time or part-time Belts, brainstorm the following:**
  - What are the benefits/risks of each choice?
  - Classify the benefits/risks as either people related or financial related.
- **If your company has selected to use full-time Black Belts, brainstorm the following:**
  - What are the potential risks as people are removed from their current positions?
  - How can we help mitigate those risks?
    - Backfilling
    - Workload shifting
  - Are there specific actions required?
- **If your company has selected to use part-time Black Belts, brainstorm the following:**
  - What are the potential risks to the success of the improvement program?
  - What are the potential risks to the success of the Black Belt?
  - How can we help mitigate these risks?
  - Are there specific actions required?

Table 13.1 Comparison Between Full-Time and Part-Time BBs

Full-Time BB	Part-Time BB
Removed from current position	Remains in current position
Backfill and succession planning is crucial	Current job continues to be done, but at lower productivity
Reports to Champion, projects can span organization	Reports to old boss, projects typically in current organization
100 percent dedicated to Six Sigma projects	50 percent dedicated to Six Sigma projects (actual data shows reality is 5 to 30 percent)
Three to four BB projects per year	One to two BB projects per year
Typically mentor five to ten Green Belts	Rarely mentor Green Belts

How Many People Do You Train? When putting together a training plan in Six Sigma, the schedule is built around what we will call waves of training. "Wave" is a Six Sigma term that is equivalent to "class." To say that you will schedule three waves of Black Belts is the same as scheduling three classes of Black Belts. HR has a leadership role in determining the number of students per wave and the number of waves of Black Belts and Green Belts per year and in what organization.

Class size for Black Belt and Green Belt training ranges from 15 to 50. Most training experts flinch at a class the size of 50. But 50 works if structured correctly. Six Sigma training includes a lot of work on laptop computers, and each module has a standup exercise built in for groups of four to six students. So, at any one time, the student is either working individually on their computer or in small group exercises. But the maximum number of students per class should be set by HR.

Recommendations for the number of students to be trained this year are found in the following table. The percentages in the table refer to the percentage of the entire population of the business sponsoring the training. So, you can plan on training 3 to 6 percent of the population as Operations Black Belts. The table shows approximate percentages for Black Belts and Green Belts and for Operations (Manufacturing), Transactional, or R&D. These percentages will give you a good start of the expected numbers to be trained to complete your training plan.

REASONABLE NUMBERS OF TRAINEES FOR THE ANNUAL FORECAST

- **Black Belts**
  - Operations — 3 to 6 percent of the Operations population.
  - Transactional — 1 to 2 percent of the Transactional population.
  - R&D — 25 percent or higher of the R&D population.
- **Green Belts**
  - Operations — 10 to 15 percent of the total population.
  - Transactional — 5 to 10 percent of the total population.
  - Design — 40 percent of the total population.

To determine a better forecast of training numbers, the following is a good HR exercise to do for each business or function. The exercise should take 45 to 60 minutes.

WORKSHOP FOR FORECASTING VOLUME OF STUDENTS TO BE TRAINED

- **Begin with organization employee count:**
  - Try to estimate numbers within Operations, Transactional, and Design processes.
- **Decide upon a target percent number for each process area and Belt level:**
  - Consider project opportunities.
  - Consider resource constraints.
- **Calculate Belt counts for process areas and Belt levels.**
- **Create a schedule for meeting goals:**
  - Two months to train GBs, four months to train BBs, and eight months to train MBBs.
  - Consider training resource constraints, costs, number per course, and ability for organization to support.

The outcome of such a workshop should be a forecasted student load by business and sponsored by HR. Table 13.2 is an example for a Black Belt training plan for a $14 billion business consisting of several business units.

Table 13.2 Annual Training Plan Example Representing the Planned Number of Black Belts (by Division)

Business	Wave 1	Wave 2	Wave 3	Wave 4	Wave 5
1. Polymers	21	22	17	23	19
2. Fluorine	5	4	2	2	2
3. Electrical	8	6	3	1	4
4. Spec Chem	4	1	2	2	1
Other					
—Carbon Tech	2				2
—MOD		1		2	
—Amorph Mtl	2	2	3	1	2
—Spec Film	2	1	1	2	
5. Engineering	3	1	2	1	1
6. JV—Asec	2	3	1		1
—UOP	3	5	5	5	8
7. R&T	3	3	5	1	1
Champions	18			1	
Asia	6				6
Europe				1	1
Total	79	49	41	42	48
Total - 1996			169		

Belt Selection Process. Black Belts and Green Belts are the engine in the Six Sigma program. The candidates for these positions should be selected carefully. I know of one large company that would only allow those people in the high-potential list into the Black Belt program. I think that's narrowing the focus of Black Belt training in that Black Belt training will create new high potentials.

A selection process should be designed to make certain the best employees are trained as Belts. The last thing a Master Black Belt who is doing the Black Belt training wants to hear in training is, "I don't know why I'm here. My manager sent me." There are several possible methods to select Belts that are dependent on the Belt level. Usually, the candidates are selected by the Deployment and Project Champions.

Poor Selection Methods

- Squeaky wheels
- Random
- Move around the poor performers

Good Selection Methods

- Manager recommendations (written)
- Formal applications

Table 13.3 shows some possible criteria for selecting Belts. The most important question is, "Can this person get the kind of results that are expected?" Before asking managers for recommendations or employees to volunteer, make certain they all know enough about the Six Sigma program, so they can make an educated choice.

Table 13.3 Criteria and Selection Methods for Black Belts, Green Belts, and Master Black Belts

	Green Belts	Black Belts	Master Black Belts
Possible criteria	Previous reviews, peer respect, departmental knowledge, and computer skills	Previous reviews, peer respect, subordinate respect, past project management success, cross-functional knowledge, technical skills, education level, and computer skills	Successful BB project completion, mentoring skills, training skills, overall business knowledge, and technical skills
Minimum selection methods	Volunteers	Management recommendations	Written applications

The following is an example agenda of an HR workshop that leads to the development of a selection process to ensure consistency across the organization. This workshop should take about 45 to 60 minutes.

WORKSHOP FOR DEVELOPING THE BELT SELECTION PROCESS

- Brainstorm potential criteria for belt selection:
  - What qualities would you like to see in your belts?
  - How might you compare one candidate to another?
- Brainstorm steps in selection process.
- Create selection process map:
  - Start with SIPOC (Supplier, Inputs, Process, Outputs, and Customer) map

This workshop should generate the selection criteria for Belt candidates, such as what is shown in the next section. This is an actual appendix of a letter I sent to all the product development Champions in my sector of AlliedSignal. The first part of the letter was requesting names of Product Development Black Belts for the first wave of training.

PRODUCT DEVELOPMENT BLACK BELTS: SOME CRITERIA AND INFORMATION

Some Thoughts, Information, and Criteria for Selecting PDBBs

Personal training and development cycle for your PDBBs:

- A four-month cycle of skills building in the customer-linked commercialization process, variation identification, and defect elimination. Each month will be comprised of four segments: "Plan/Train/Apply/Review."

  The "Plan/Train" segments are in the first week of each month.

  The "Apply" segment is back in the Product Development Black Belt's business area working on his/her specific objective.

  The "Review" segment is covered with "Apply" at the start of the next month with "Plan."
- The four months will focus our Product Development Black Belts on applying advanced tools within the Customer Linked Commercialization process.

The Product Development Black Belt Training Cycle [WAVE 1] starts on April *15TH* at the Omni Hotel in Richmond, Virginia. Detailed information has been sent out to you and your PDBBs.

PRODUCT DEVELOPMENT BLACK BELTS

According to Dr. Mikel Harry, an early leader in Six Sigma deployments, the most successful candidates have typically been individuals who have the potential to realize a synergistic proficiency between their respective discipline and the strategies, tactics, and tools of variation and defect elimination (Six Sigma).

- **The suggested characteristics for candidates include** resiliency, driven by a purpose, leads from knowledge, enjoys and is good at hands-on involvement, acts on fact, and can function as consultant, leader, or gladiator as appropriate.
- **The individuals that you select** should be expected to become a participant in the operational directing of your business. They should be able to help surface the most significant opportunities for improving your new product/process development process, new product quality, and new process consistency. You must be willing and able to put high expectations on them.
- **The profile** will probably include at least a four-year technical/scientific degree or its equivalent, new product and/or process development experience, and *strong* personal computer skills. Desired additional expertise includes experience in leading teams, training in basic statistics and/or control techniques, and effective presentation skills.

You can see that strong HR support will facilitate the launching of Six Sigma by sponsoring critical HR workshops soon after Executive Team training. Here are some of the considerations in supporting your organizations in selecting the right people.

Actions for Identifying the Right People

1. Understand the roles and skill sets needed in Six Sigma.
2. Identify available talent internally.
3. Work with management to free up existing talent and hire/backfill or delegate responsibilities as needed.
4. Coach/counsel those selected on the impact of the opportunity.
5. Aid/coach in adapting to the new role and giving up old responsibilities.

Developing Job Descriptions

Early in the Six Sigma deployment planning stage, HR will effectively support the deployment by understanding the new roles and requiring and facilitating the development of job descriptions for the new positions. New descriptions will be required for Master Black Belts, Black Belts, and Green Belts. The following is an example of a position profile for Green Belts. These position profiles drive consistency in selection criteria, as well as deploy the vision of the program. These position profiles act as catalysts for the upcoming change and also demystify the cultural change about to occur.

POSITION PROFILE

POSITION TITLE: SIX SIGMA GREEN BELT SALARY GRADE: TBD

Key Responsibilities

- On a part-time basis, lead and manage improvement projects in own functional area as assigned by Champions and Sponsors.
- On a part-time basis, support Black Belt projects.
- Coach improvement project team members on the improvement process and corresponding quality tools.
- Make regular reports to management on improvement project progress, barriers, and issues.
- Participate in potential project selection and analysis; advise BU staff on prioritization of potential projects.

Most Critical Skill Dimensions

- **Functional/Technical**
  - User-level skill in basic product/process problem-solving techniques.
  - Operational experience of meeting objectives.
  - Two-year technical or professional education, or equivalent experience.
  - Microsoft Word, Excel, PowerPoint, and Project mid-level user skills.
  - Written and oral presentation skills.

- **Leadership**
  - **Think strategically:**
    - From experience and knowledge, considers a broad range of internal and external factors when solving problems and making decisions.
    - Identifies critical, high pay-off strategies and prioritizes team efforts accordingly.
    - Recognizes strategic opportunities for success; adjusts actions and decisions for focus on critical strategic issues.
  - **Analyze issues:**
    - Gathers relevant information systematically.
    - Grasps complexities and perceives relationships among problems or issues.
    - Seeks input from others; uses accurate logic and data in analyses.
  - **Persistently champion change:**
    - Challenges the status quo and champions new initiatives.
    - Acts as a catalyst for change and stimulates others to change.
    - Paves the way for needed changes.
    - Manages implementation effectively.
- **Management**
  - **Foster teamwork:**
    - Builds effective teams committed to organizational goals.
    - Fosters collaboration among team members and among teams.
  - **Manage execution:**
    - Assigns responsibilities.
    - Delegates to and empowers others.
    - Removes obstacles.
    - Allows for and contributes needed resources.
    - Coordinates work efforts when necessary; monitors progress.
  - **Develop systems and processes:**
    - Identifies and implements effective processes and procedures for accomplishing work.
  - **Commit to quality:**
    - Emphasizes the need to deliver quality products and/or services.

    - Defines standards for quality and evaluates products, processes, and/or services against those standards.
    - Manages quality.
- **Coach**
  - **Foster open communication:**
    - Creates an atmosphere in which timely and high-quality information flows smoothly between self and others.
    - Encourages the open expression of ideas and opinions.
  - **Training:**
  - Ability to assess training needs of team members and determine appropriate training/development to transfer knowledge, skills, and abilities.

The following list presents a summary of the roles of Champions, Master Black Belts, Black Belts, and Green Belts. Workshops with the Six Sigma Steering Teams should flesh out the remaining parts of the position profiles. Some good work on defining the Six Sigma infrastructure operationally will be like oil on the gears of the deployment. The vision of the program will be operationally defined and clarified.

Champion

- Break Barriers
- Reward and Recognition
- Project Selection
- Monitors Project Progress
- Identify/Request Candidates for Training
- Recommends/Prioritizes BB Assignments (Time)

Master Black Belt

- Mentor BBs/GBs
- Training of GBs/BBs—Create and/or Extend Material
- Institutionalize Six Sigma Philosophy
- Leadership

- Communicator/Facilitator at All Levels of Management
- Identify and Communicate Opportunities
- Identify and Communicate Barriers
- Speak Language of Management
- Innovative Tool User

Black Belt

- Understanding "The Metrics"
- Team Player/Assistant
- Show "Extraordinary" Sense
- Expert in Applying Tools
- Mentors GBS/YBS
- Bias for Action
- Team Facilitation Skills
- Drives "Big Money" Projects to Completion
- Technical Background
- Proficient with Computers

Green Belt

- Understands and Applies Breakthrough Strategies—DMAIC
- Works with Workgroup Teams
- Works with BBs on Large Projects
- Basic Computer Skills

Succession Planning. When a person is selected for Black Belt training, what their role is during and after training depends on the full-time and part-time decision. All Black Belt students show up to the first class with a chartered project in hand, so project work starts on the first day of class. If the student is to be full time on the project, then the manager, the Belt, and their HR associate should be required to create a succession plan to cover his or her previous duties.

The Black Belt's position should be backfilled or existing responsibilities reallocated as soon before the training as possible. I remember visiting one of my chemical plants and the Six Sigma Steering Team for the plant was explaining why their Black Belts could not be full time on project work. Then, suddenly, a Black Belt chimed in. "I'll tell you why I'm not full time on project work. It's because I'm doing a lot of things that a chemical engineer shouldn't be doing in the first place." And she was right.

The Steering Team admitted they could hire a few college interns to cover some of the Belt's former responsibilities. The Steering Team had a new plant manager, who joined the company from another chemical company that had a Six Sigma program. He mentioned in a side conversation, "I know these people are worth $1,000,000 a year to me. Why shouldn't I reallocate some of their responsibility, or even hire a few more to add to the ones I already have?" I was thrilled to have someone on the team who saw the value of the program.

Succession planning already in process should consider the addition of new future leaders to the succession pool. Although most Black and Green Belts enjoy the technical aspect of Six Sigma, most are interested in moving up the leadership ladder as quickly as possible. Most want to run a factory or a business by the end of their careers.

Consideration of the new leadership skills, developed as a Black Belt, should help drive the traditional succession planning. If the HR function helps muster the support for Six Sigma being formally considered part of the leadership development roadmap of your company, recruiting new Black Belts and Green Belts will be no problem.

Just as in the case of GE, once Welch sent out the edict that all future leaders would be certified Green Belts, the demand for Green Belt training was enormous. GE trained over 64,000 Green Belts in two years. Because the Belts are learning dynamic, real-time leadership skills while completing high-value projects, it makes sense to tie those efforts into the leadership career roadmap.

Certification Requirements. A standard feature of Six Sigma deployments is the development and implementation of certification requirements for the Belt levels. Some companies even certify the Champions. The HR function has the opportunity to lead the development of the company's certification requirements and to archive the list of those certified. There are no industry-standard certifications, though some organizations offer that service. It is best for each company to develop the certification standards that best fits their culture. Certification requirements will vary from company to company and by Belt level. Table 13.4 summarizes the differences in certification among the Belts.

Typically, certification criteria includes

- Attending all training.
- Successfully completing and reviewing projects.

- Examination—this has become more common for GBs and BBs. Not typically applicable for MBBs.
- Minimum savings—highly dependent on average project savings; varies heavily from company to company. Must decide whether soft savings should count toward certification.
- Training of others—teams, GBs, BBs, or Champions.
- Coaching and mentoring of others—applicable for BBs and MBBs.

Table 13.4 Certification Guidelines for Different Belt Levels

	Green Belts	Black Belts	Master Black Belts
Attended training	Two weeks	Four weeks/Five weeks (DFSS)	Five weeks and two electives
Projects	One to two completed	Two to four completed	Four or more completed
Minimum savings	Company/department dependent	Company dependent	Company dependent
Exam	Company dependent	Company dependent	None
Training others	Have trained team in tool usage	Have trained some GB modules	Can train all BB, GB, and Champion curriculum modules
Coaching others	N/A	Coached X number of GBs to project completion	Coached many GBs and BBs to project completion

The following is a short exercise that the HR function can put together with their Six Sigma Steering Team to draft certification requirements.

Certification Exercise

With each Belt level, discuss the following criteria:

- **Number of Completed Projects**
  - How many projects do you expect GBs to complete successfully?
- **Savings $**
  - How much do you expect an average project to return (consider ops, trans, and design)?

    - Do you want to put a minimum amount of savings for the projects? Should this include both hard and soft dollars? What are the risks/benefits here?
- **Exam**
    - Are you planning on having an exam? What are the risks/benefits?
    - When/where/how will you administer?
- **Training Others**
    - Will Belts be required to train others?
    - Entire curriculums or just portions?
- **Coaching/Mentoring Others**
    - Will Belts be coaching other Belts?
    - Will it be GB to BB? Or experienced BB to BB, and so on?
    - What do you want the mentoring hierarchy to look like?
    - What is the criteria for successful coaching?

Do as much of this as you can. Assign action items to discuss further or set up plans.

RECOGNITION AND REWARDS

Six Sigma deployments have a magnitude of culture change associated with it. Therefore, a clear, aggressive, and innovative rewards and recognition (R&R) program defined early in the deployment will give the deployment a great start. Every company has its own R&R system, and that is a good place to start with the new Six Sigma R&R system.

Because of the nature of Six Sigma, some innovative additions and modifications of the current system will be in order. As I previously mentioned in an earlier chapter, the engineered materials sector of AlliedSignal launched an entirely new R&R system called Winning Together. The system was sector-wide and included the entire population. As a company, AlliedSignal started a corporate-wide R&R program to reward outstanding applications of Six Sigma across the company.

The originator of Six Sigma, Motorola, established a fine program called the Total Customer Satisfaction (TCS) program. The TCS program sponsored a huge conference allowing the outstanding team within each business to present their Six Sigma projects and compete for corporate-level prizes. The team members were treated like royalty, being driven around in limousines and staying at the finest hotels. The program was highly effective in driving and rewarding the right behaviors.

Depending upon your company's existing culture and vision of its future culture, a very robust recognition program can often mitigate the need for rewards. The best rewards and recognition plans treat the impending improvement program as part of annual performance planning process. Recognition in the form of career development is extremely effective because there is career advancement and recognition of talents and money (raises and stock options) involved.

Rewards in terms of compensation adjustments (annual bonuses, or raise/promotion process) are the strong part of every successful Six Sigma deployment. So, the combination of recognition through career advancement and reward through aggressive compensation becomes the dynamic duo.

A risk in Six Sigma deployments is that too much attention is put on the Belt (Master, Black, and Green). The R&R structure should touch all Six Sigma participants, not just the Belts. The new R&R systems should be primarily given based on results, not just participation. The following table suggests a few simple forms of recognition. The section immediately following outlines a potential workshop to develop methods of recognition for Six Sigma.

SUGGESTIONS FOR RECOGNITION ACTIONS

Suggested Workshop for Developing Recognition Methods

- **Purpose:** To identify opportunities for recognition within your improvement program.
- **Exercise:**
  - List current recognition avenues in your company and improvement program.
  - How might you use these to provide recognition to improvement teams?
  - Brainstorm additional recognition ideas that you may use.
  - List benefits and concerns for each idea.
  - Create an action plan around next steps to have recognition program in place prior to first wave of belts.
- **Time:** 60 minutes.

Sincere thank yous are always great rewards. Table 13.5 presents possible reward methods, with notes. Along with recognition, there are rewards. The following list addresses rewards and Six Sigma:

- **Rewards can be recognition based:**
  - Plaques, certificates

- **Career oriented:**
  - Conferences
  - Competitions
- **Promotion potential:**
  - Exposure to upper management
- **Monetary:**
  - Gift certificates
  - Certification bonuses
  - Annual performance raises or bonuses
  - Best project competition awards
  - Savings sharing
- **SBTI has found that the best rewards are:**
  - Recognition
  - Learning opportunities
  - Career growth
  - Competitive compensation tied to Belt successes
- **We do not recommend sharing bonuses with the Belt and/or teams:**
  - Creates a "WIIFM" environment
  - Drives competitive behavior between Belts and teams

Table 13.5 Other Reward Methods with Notes

Possible Rewards	Notes
Budgets for team expenses, such as lunches and outings	Should have minimum budget for Belt for some team expenses
Team gift certificates	Successful project completion
Belt conferences	Belt conferences sharing best practices and continuing education; can tie to competition for best projects
Project competitions	Annual "best project" competition; can have criteria, rewards, and so on
Annual performance reviews	Raises, bonuses based on improvement and goal completion

Possible Rewards	Notes
Certification plaques	Given upon completion of certification to Belts
Certification bonuses	Upon completion of certification; often part of agreeing to Belt training
Project savings sharing bonuses	For Belts and/or team members upon project completion; usually percentage of project

ACTUAL EXAMPLES OF REWARD AND RECOGNITION

Motorola

- Total Customer Satisfaction Competition: A formal company-wide competition, focused on impact of quality teams. TCS Competition is a big deal at Motorola!
- Competition started at plant level and worked up to corporate level.
- Corporate competition sites included Paris, Hawaii, Singapore, Bali, and so on.
- Reward was the trip and opportunity to participate.
- Site reinforced the global nature of the business.
- Similar competition launched in AlliedSignal with equal success.
- Highly recommended.

General Electric

- 40 percent of annual bonus related to Six Sigma activities.
- Green Belt certification required for promotion.

"With Six Sigma permeating much of what we do, it will be unthinkable to hire, promote, or tolerate those who cannot, or will not, commit to this way of work."
—Jack Welch, CEO, General Electric (USA Today, February, 1998)

Company A, B, and C

- **Company A**
  - Week 4 celebrations and stock options for BBs.
- **Company B**
  - Nothing—BB projects part of everyday job.
- **Company C**
  - BB graduation dinner with high-level executives.
  - $500 to $5,000 for certification depending on impact.
  - Gifts and stock options for teams.
  - MBB received stock ($25K value) on certification.
  - These stock awards were reduced later in the initiative.

Recognition and Rewards for Black Belts: Company D

- **Company D**
  - Black Belt shirts given the second week to acknowledge participation.
  - Technical conferences for presenting project results, sometimes to the next class of BBs.
  - Much bigger pay raises—up to 2x of non-BB peers.
  - Ability to nominate key team members for $500 to $1,000 bonuses.
  - Individual write-ups in company newsletter with "blow-up" of article and picture to hang in office.
- **Stepwise Recognition**
  - BB shirt during training.
  - Dinner with Management with gift check for completing training.
  - Bonus at certification commensurate with results; very nice trophy or plaque included with certification certificate.

Company A Approaches Career Path and Compensation

- **Career Path/Title**
  - Black Belts are full time and have been given a title with the word "Black Belt" in it to reflect this role.

    - They will stay at their current "rank" or "band" in the organization; that is, there is no common rank/band for these folks. Some came in at the Director level and will remain at the Director level; others are "junior" engineers and remain at that level.
    - They are expected to remain in the role for two to four years.
    - However, the president of the largest division made a commitment to allow any of them to return to their prior role at any time if they don't feel that the Black Belt role is right for them.
- **Compensation**
    - Base compensation does not change with the move to Black Belt.
    - Any bonus potential to which they were previously entitled remains intact.
    - All Black Belts receive 500 to 1,000 stock options at the completion of training (actual number still TBD).
    - All Black Belts entitled to performance bonus based on the results of their projects—from 0 percent for no results to about 10 percent of salary—for meeting targets to greater levels if they blow the doors off their project goals.
    - All Black Belts are "enrolled" in their EBITDA-based bonus program, which is tied to the performance of the company overall (this is incremental for most of them).

BELT RETENTION AND CAREER PLANNING

Belt Retention. One of the unfortunate aspects of deploying Six Sigma is that there is now a premium on Master Black Belts and Black Belts in the corporate world today. Therefore, the Belts you train will be a highly sought-after asset in the market. Your Black Belt students will start getting calls from headhunters before they're even finished with training. Entire headhunting firms have been created that do nothing but locate Six Sigma resources. My son went through my Master Black Belt training and he has had no problem finding great jobs.

What does this mean to you? You, as an HR professional, need to put retention actions to the forefront. Retention will be an issue. My sector in AlliedSignal focused heavily on retention, and out of 350 Black Belts training in two years, we only lost two, while the other sectors were losing Black Belts in high numbers. Table 13.6 shows some retention risks with potential solutions.

Table 13.6 Retention Risks and Potential Solutions

Retention Risk	Potential Solution
Higher pay opportunities	Pay competitively in the marketplace and offer the Belt growth potential.
Perceived promotions	Give Belts a clear career path; Black Belts should have a two- to three-year position that grows into greater career opportunities.
New challenges	Variety in projects; projects in other areas, such as facilities divisions; opportunities to train folks.
Lack of support	Good Champion screening and training.
Mixed priorities	Focused deployment with clear goals and objectives for all.

Even the best rewards and recognition program cannot prevent this, but people are motivated to work on tasks/projects that will be beneficial to their career goals. If you do a good job at keeping them on a learning curve with a *clear career path*, they are less likely to become dissatisfied. Black Belts and Green Belts are relatively easy to satisfy. You let them work multiple high-impact projects, provide them some good continuous education opportunities and some great recognition events, slide some impressive financial rewards in, and you will keep your Belts. They are special and treating them special (based on results) makes sense. They're the future of your company.

Career Planning. Establish distinct links between success as a Black or Green Belt and the furthering of their professional career. Performance objectives should be linked to the project goals and reflect the Belt's chartered projects. Champion and Business Leader/Plant Manager performance objectives should reflect project goals as well. Each Belt should have a clear career development plan established during the first year of their assignment. This should include the following:

- Growth opportunities as a Belt—other areas or facilities; training of other Belts or peers.
- Continuing education opportunities—additional training in advanced tools; post Belt role.
- Career advancement plans directed at positions.

The next exhibit shows a recommended workshop that HR professionals have held to directly address the issue of retention and career planning. This workshop should only last 60 minutes but should save the company a lot of lost resources by keeping highly qualified Belts around.

WORKSHOP ADDRESSING RETENTION OF BELTS

- **Purpose:** To identify retention risks and mitigate them.
- **Exercise:**
  - Your ultimate goal is Belt retention.
  - Brainstorm reasons why Belts may decide to depart.
  - Logically group the reasons.
  - Highlight reasons you can help mitigate.
  - Identify steps and action items to assist risk mitigation.
- **Time:** 60 minutes.

Because Six Sigma relies on metrics to determine how good the program is doing, establishing HR metrics such as percentage of students certified, percentage of Belts that have left the company, and tracking the career progression of certified Belts is essential.

SIX SIGMA DEPLOYMENT HR SUPPORT

There are multiple ways in which HR can support a deployment. Facilitating the communication plan for Six Sigma is necessary to aid in accelerating support to help the Belts effectively lead teams. There will also be a requirement to write a lot of letters to the troops by the leadership team. HR support in generating those letters will streamline the deployment.

Communication Plan. Once the deployment is designed, communication should begin. A Communication workshop should be held. The following questions should be answered and the plan crafted. See Chapter 12, "Communicating the Six Sigma Program Expectations and Metrics," for more information on communication planning.

- Who do you need to communicate with?
- What are their main concerns?
- What is the best medium?
  - Newsletter, email, v-mail
  - Town hall or all hands, small groups
  - Memo/letter, video, web page
  - One-on-ones
- How often should they be communicated with?

Team Facilitation. With a Six Sigma deployment, hundreds of teams will be launched, and Black Belts and Green Belts will be new to this role. Because the projects are important, grooming your Belts to facilitate teams will be essential. In fact, this should be directly included in the Belt training programs. Focusing on team staffing is of concern. People who have a reputation of getting results often are overallocated to teams. HR should track the percentage of time that is spent by team members in teams. This is best centrally done, coordinating all team-based projects within the organization. The skills that should be deployed along with Six Sigma are team leadership support, facilitation skills, and conflict resolution.

Draft Deployment Letters. Considering the amount of communication necessary for any change initiative, support in drafting some of the initial communications is most welcome. Following is a sample letter from a company president concerning the impending Six Sigma deployment.

SPECIALTY CHEMICALS VISION

Date January 10, 2001
To: Site Champions
From President
Subject: Process Metrics for Six Sigma Performance

Specialty Chemicals will become the number one or two in the businesses in which it operates. The number one or two positions should be in the areas of operations, customer value creation, global share, and profitability. Further, it will be a company that attracts, retains, and develops the best people in the industry. In terms of Operational Excellence, Specialty Chemicals will operate its facilities in a safe and environmentally responsible manner and produce, at lowest-in-industry costs, the high-quality products our customers expect. We should continually drive productivity and improvement in our global production assets.

Our History of Success

Specialty Chemicals has been itself a history of success in terms of manufacturing excellence, thanks to the good work, talent, and commitment of its people. Among other successes, we have built and started up new units all over the world, have operated and expanded our units in a sound manner, and have implemented cost reduction projects. However, this has not been enough in terms of keeping the pace with our competitors. New challenges are in front of us, and I am sure, like in the past times, we could deal with them in an effective and expeditious manner.

Our Challenge

Maximization of current capital and labor assets, as well as the best utilization of materials and energy, have become critical bases of competition in most manufacturing industries around the world. Superior and continuously improving operational performance is rapidly moving from a *Strategic Advantage* to a *Strategic Requirement.* The production of substandard products either in *quantity or quality and yields* is a clear waste of capital and labor assets, as well as money, and detracts directly from the competitiveness of a business.

We must each commit to improve our Operational Performance by 7 percent each year over the next three years. An improvement of this magnitude does not occur from continuing to *do the same old things in the same old ways. . .*or even *from doing them a little better or faster.* From today forward, we must focus our decisions, our actions, and our people on the journey to Premier Operational Performance. *Six Sigma is a process that will enhance our current initiatives by means of incorporating into them powerful statistical tools. It is focused not only on further reducing costs, but also by increasing revenue in some cases. The main attribute of Six Sigma is to reduce variability in both operational and administrative processes, thus increasing productivity.*

Your First Assignment*:* Each manufacturing plant leader, improvement Champion, and staff should focus on their most important strategic products and processes during the *next* 10 days. At each plant, the processes representing the majority of your business for 2001 should be examined utilizing the Six Sigma Process Maps and the Six Sigma metrics of Rolled Throughput Yield (RTY), Cost Of Poor Quality (COPQ), and Capacity Productivity (C-P). I would like to have your best inputs sent to my administrator by *Thursday, January 25th*. This effort will not be easy and must have your personal commitment, because it is critical to start our journey with the right first steps.

In mid-February, I will meet with the Manufacturing Council and our Business Leaders to review each of your inputs. To make this easier and more uniform across our many sites, we are including in this memo blank maps, the appropriate metric definitions, and your preparation

> instructions. Also, please find attached to this message electronic copies of each of these items for your use.
>
> I would like to provide assistance, so please contact either Anthony, Less, or James if you need help in this important effort.
>
> Thank you for already playing a critical role in the future of Specialty Chemicals,
>
> Richard

There are a myriad of other issues on which HR can work. For example, if your company has a strong union, planning on how to address that issue early is very important. There's also the whole change management arena in which HR folks have experience. Look at Six Sigma as a chance to take an HR leadership role. If the top Six Sigma companies are benchmarked, you will find a strong HR effort behind the deployment. The senior HR person will be a member of the Six Sigma Steering Team, so there will be ample opportunity to do great things early in the deployment.

14 Defining the Software Infrastructure: Tracking the Program and Projects

With Dino Hernandez and Instantis, Inc.

It's one thing to launch a corporate-wide change initiative like Six Sigma, but it's totally another thing to evaluate the return on investment of the program across the corporation. Larry Bossidy, Jack Welch, and Jim McNerney made no question that the purpose of Six Sigma was to change the company and get measurable financial and business results.

An advantage you have over the early icons of Six Sigma like Larry Bossidy and his team at AlliedSignal is the access to a rich selection of information and process management resources to accelerate both the results and expansion of your deployment. These new tools offer the first step in managing the enterprise. This chapter covers the following general topics, with detail for each:

- Enterprise Management Fundamentals
- Enterprise Solutions
    - Getting Started
    - Fundamentals
    - Available Packages
    - Desktop Solutions
- Planning Solutions
- Executing Solutions
- Tuning and Expanding Solutions

In the early days of the AlliedSignal engineered materials Six Sigma deployment, all program tracking (number of Black Belts trained and certified, financial results of completed projects, and the performance of each business) was done by Excel spreadsheets and other simple software applications like Microsoft Project. I soon fell behind the power curve, and tracking became extremely time consuming and difficult.

We soon grew our own Internet-based software application specifically to track Six Sigma actions. It was called Viper X, named after the Six Sigma mascot, the Viper sports car. Now there are numerous options in the market that will allow you a turnkey enterprise-wide system. Tyco implemented an enterprise system in a very short period of time, as have several other large companies. In fact, Six Sigma consulting companies have scrambled to provide their own solution to the Six Sigma deployment market.

Six Sigma software solutions available today have a number of overlapping features and capabilities, but there is a clear distinction between networked, enterprise management applications and desktop project execution tools. Deployment Champions must rely on a portfolio of software solutions that allow for the right balance of personal flexibility regarding desktop project-level tools, but without sacrificing the capability to roll up results, track performance, and set strategic priorities within a centralized, networked management application.

For example, Table 14.1 shows a rollup of one of my early deployments. This report presented some cumulative statistics of the first three years of the deployment (1995, 1996, and 1997). The report included cumulative financial impact, operational metrics, and numbers of Belts trained. This report was put together using desktop applications—mostly Excel spreadsheets. A lot of time and energy went into this report. But, with the enterprise systems available today, I could have generated this report with a few keystrokes and mouse clicks.

Table 14.1 Longitudinal Statistics for an Enterprise-wide Six Sigma Initiative

Business Metrics	**1995**	**1996**	**1997**
Operating Income	$90M	$110M	$130M–$140M
Net Income	$54M	$61M	$78M–$84M
OE Metrics	**1995**	**1996**	**1997**
Rolled Yield	82%	85%	88%
COPQ	12.5%	9.7%	8.3%
Capacity-Productivity	1.00	1.10	1.20

Training Metrics	1995	1996	1997
Master Black Belts	6	14	20
Black Belts	170	308	450
Product Development Black Belts	0	65	165
Green Belts	222	522	822
DFSS Green Belts	20	75	275
Yellow Belts	0	125	1,000
Total	**418**	**1,109**	**2,732**

Teams that design their information systems well and employ an effective balance of project tools and program management solutions are able to do the following:

- Complete projects more effectively.
- Apply tools more consistently.
- Track and report more immediately.
- Collaborate more easily.
- Realize greater program results sooner.

ENTERPRISE MANAGEMENT SOLUTION FUNDAMENTALS

Any initiative that has the capability to efficiently track, manage, and align project-level activity to corporate-level strategic objectives and business priorities has a high probability of success. Different organizations achieve differing levels of visibility and control over their initiatives directly proportional to the investment they commit to the systems required to support their teams.

At the beginning, the initiative leaders use software to simply track their Six Sigma programs and strategies in spreadsheets, personal project tracking software, and other desktop tools. However, these standalone, desktop approaches cannot meet the needs of a team with managing more than a handful of Six Sigma projects or strategies.

These shared local files and ad hoc communications flowing between the team members to track results quickly breaks down from manual errors and conflicting versions of the information as the number of projects and team members increase. You can imagine the kind of manual work required to put together this report on operational results for a single division of a company (see Table 14.2). You would be begging for an automated enterprise system.

Table 14.2 Longitudinal Statistics for an Enterprise-wide Six Sigma Initiative

Key Metrics	1994 Baseline	1995 Goal	First Quarter Goal	First Quarter Actual	Second Quarter Goal	Second Quarter Actual
RTY	77%	82%	79%	80%	81.3%	79.4%
COPQ	19%	15%	17%	17%	16%	14.5%
Capacity	1	1.14	1.09	1.08	1.13	1.13

A next-level solution is to establish a basic data-sharing infrastructure. These solutions are usually built on a general-purpose platform like Access or Project Server/Sharepoint, or are borrowed from a corporate database already in use for other systems. These internally configured solutions provide some level of centralization and a common repository of project information.

They can be expensive to create and are difficult to maintain, and are therefore normally limited to a few specialized features. For a business that is growing rapidly or expanding the scope of its Six Sigma initiatives, systems of this type usually hit their envelope of capability quickly.

Specifically developed, enterprise project portfolio software solutions provide a central database of Six Sigma project and program information, offer a common web-based interface for users, and enforce workflows for the following:

- Project selection
- Work assignments
- Progress reports
- Tracking of metrics and financials for each project across the enterprise

With this level of infrastructure in place, the team benefits from real-time access to its entire portfolio of Six Sigma project information and reporting for review and analysis. These systems also perform a valuable compliance role by assuring best uses of the intended tools and defined roadmaps supported by the system. However, these software packages are focused on the execution side of Six Sigma and overlook the need for aligning the project portfolio to the strategic priorities of the company.

Beyond the core enterprise project portfolio capabilities, leading-edge solutions track the following:

- Day-to-day project activity and tools.
- Measure how well the entire portfolio of projects (and each project individually) align with the key objectives:
  - Objectives defined within the organization's overall corporate strategy.
  - When deployed, these solutions effectively enable an organization's strategic plan and the Six Sigma project portfolio to become one and the same.

ENTERPRISE SOLUTIONS: GETTING STARTED

Dedicated project information systems are required to transform operational and strategic alignment from a series of loosely coordinated team meetings, training sessions, or executive retreats into an essential element of day-to-day business. With these systems in place, every member of the organization, regardless of location or role, can literally remain "on the same page" as to the overall expectation and the progress to date required for the team and for the full portfolio of projects underway.

Any software solution or solution suite designed to address these integration challenges must provide some combination of all of the following attributes:

- Capability to aggregate projects and project data aligned to specific strategic initiatives.
- Project selection capabilities to assign, prioritize, and measure projects based on their specific strategic contributions.
- Universal web-based access to avoid IT inconsistencies and to assure accessibility.
- Flexibility to coordinate any variety of process initiatives or desktop applications that are in use within the individual business units.

In the implementation of this solution, the deployment managers must work aggressively to

- Assure full compliance to the basic day-to-day workflow and reporting infrastructure contributing to the overall metrics of the initiative.
- Effectively integrate (or mirror) the system reports to the metrics and reports tracked by the executive managing the team.

- Maintain and continuously manage the knowledge base and post-project reporting so that effective results and learning stemming from earlier project efforts can be more widely coordinated or more widely duplicated in other projects.

ENTERPRISE SOLUTION REQUIREMENTS

A popular Six Sigma maxim is "you can't manage what you can't measure." This is especially true for Enterprise Six Sigma initiatives. At the executive level, the Executive Team will have difficulty in gaining accurate and timely insights into the workings of a global portfolio of projects and project teams. To make effective and timely leadership decisions and strategic suggestions becomes virtually impossible without insight into the daily workings of your teams.

For example, 3M's Six Sigma deployment at one point consisted of literally thousands of projects and project teams across the company. Without an effective tracking, measurement, and alignment solution, the 3M executive could not effectively do his/her job as a leader, manager, or strategist.

Early Six Sigma deployment teams who first lead the charge to migrate their Six Sigma initiatives from project by project efforts into overall strategic efforts made substantial investments to create internally built solutions to manage the tracking, alignment, and collaboration within their teams.

Today's turnkey solutions from companies like Instantis, PowerSteering, and iSolutions offer not only proven systems that are ready to run but also offer workflows and best practices refined by the input of their large communities of Six Sigma customers who use the software each day. Effectiveness of a solution needs to address the full lifecycle of a Six Sigma project and full scope of collaboration required within the initiative team. The essential capabilities include the following:

- Alignment to strategic objectives.
- Project leadership:
  - Collect project ideas and develop into projects.
  - Tracking of project team, tools, and deadlines.
  - Alignment and validation of project financials.
  - Allocation of teams and resources.
- Enterprise strategic leadership:
  - Track projects individually or as a portfolio.

- Tracking and reporting of all program results.
- Identification and sharing of best practices.

Alignment to Strategic Objectives. If a project is completed and aligns to the overall business goals, does it still add value? Alignment to the overall goals of the business is essential for the success and buy-in for any process improvement initiative. To ensure strategic alignment all the way through the entire project portfolio, an enterprise solution has the capability to

- Establish strategic hierarchies as they cascade across an organization.
- Track projects and results as they align to each defined strategy.
- Communicate strategic expectations across the organization to focus project selection and prioritize resource commitments.

Figure 14.1 shows a partial screen of just one strategy: health, safety, and environmental. There would be screenshots for each strategy and strategic metrics.

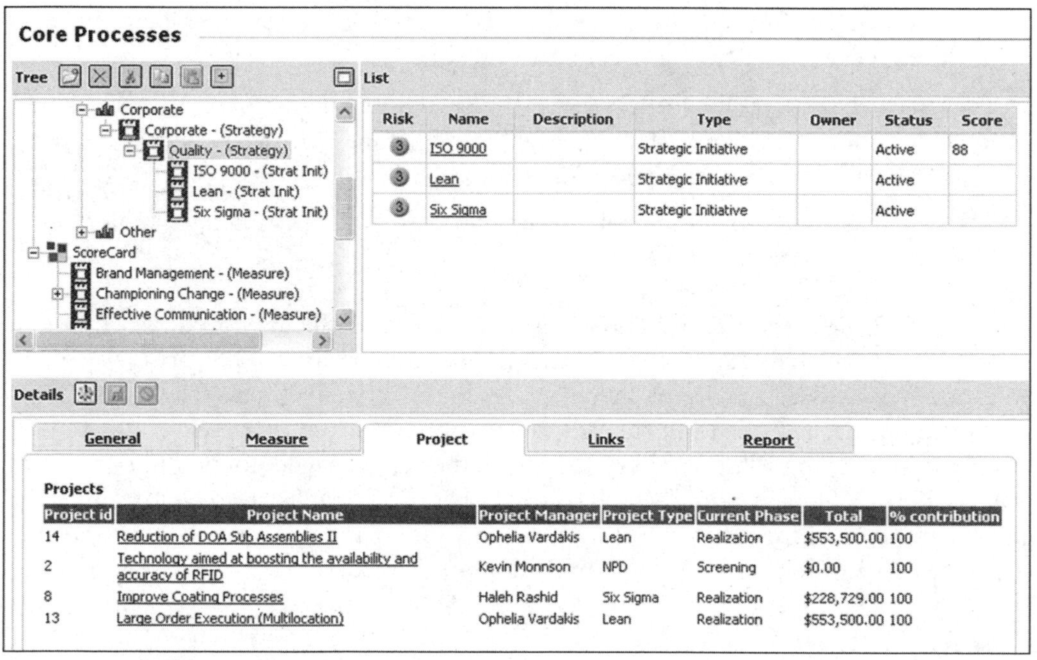

Image courtesy of Instantis, Inc.

Figure 14.1 Sample screenshot of strategic actions for the safety and environment strategic goals.

Project Leadership: Collect Project Ideas and Develop into Projects. A successful, dynamic, and widely embraced Six Sigma initiative requires a steady diet of good ideas from across the Six Sigma team and the organization as a whole. To make this happen, enterprise solutions require (1) an open door for new ideas and (2) an effective process for how they are assigned and evaluated within the team. By working from a shared enterprise system, each idea can be passed to differing contributors or other members of the team to be further filtered, developed, and better defined.

Because the act of identifying, prioritizing, and selecting high-impact projects is the soul of Six Sigma, placing these actions into a systematic process is true to the form of Six Sigma. Key to this idea of the development and selection process is instilling evaluation criteria to specifically address the strategic potential of any idea relative to the overall goals of the organization.

From an executive's perspective, an enterprise system allows the team to accept a far more aggressive flow of ideas and handle them easily through an automated and empirical process to make better project choices. The enterprise system should automate the acts described in Chapter 9, "Committing to Project Selection, Prioritization, and Chartering," which addressed project selection. Critical features include the following:

- Easily accessible Internet-based portal to collect ideas from team members, business units, and even partners, customers, and vendors.
- Scoring criteria for both business and strategic value of the idea (see Chapter 9).
- Collaborative capability to allow multiple team members from across the organization to each contribute to single project opportunity.
- Workflow to structure the contributions and approval processes required as an idea evolves into a project.

Tracking of Project Team, Tools and Deadlines. Managing the day-to-day execution of a project is the bread and butter of any performance improvement enterprise system. A by-product of a Six Sigma deployment includes the effective tools and roadmaps for completing projects. If the team and its systems cannot execute effectively at the project level, any attempt to roll up results or to expand the initiative will not be valid. If each of 500 or 1,000 Six Sigma projects would be tracked as shown in Figure 14.2, imagine trying to keep track of the progress of the entire portfolio manually. How attractive is it to just use a few mouse clicks to get the information you needed?

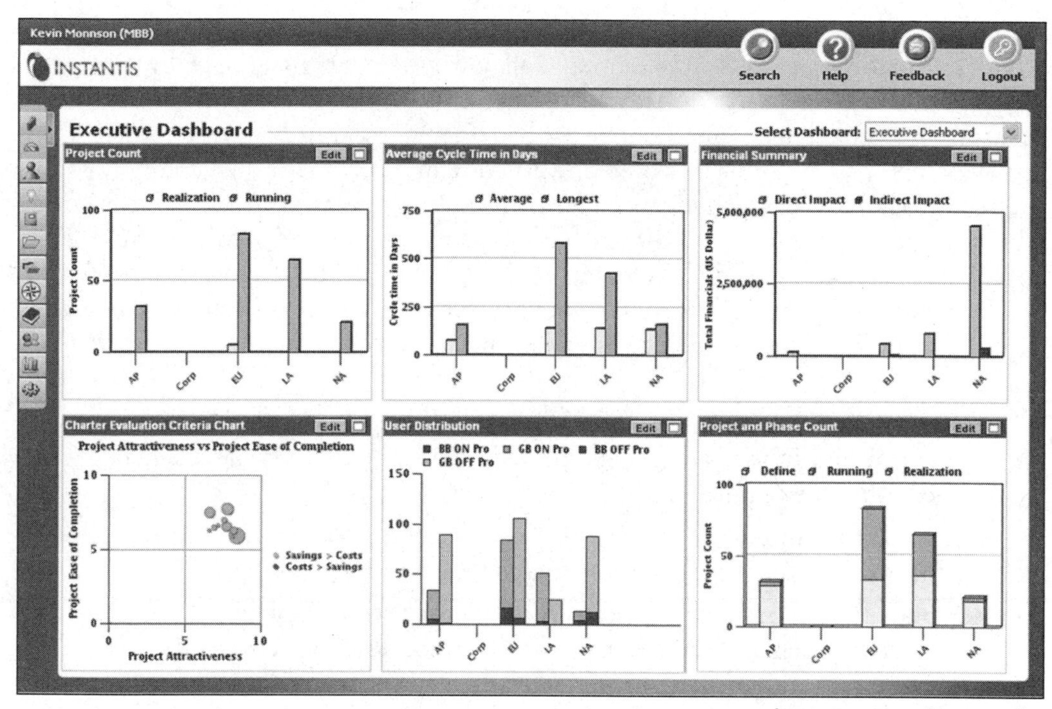

Image courtesy of Instantis, Inc.

Figure 14.2 A partial screenshot of the progress of an actual single Six Sigma project.

To add some challenge to this requirement, project teams (like their businesses) are widely distributed and come from many areas of expertise. The enterprise application platform provides each team member and their leadership a shared touchstone to collaborate and coordinate on every phase of the project. The application must be able to provide every team member with not only their expected assignments and due dates but also the recommended tools and any supporting context or information they can leverage to best execute their contribution to the team. Expected capabilities should include the following:

- Flexibility to tune workflow and project roadmap configurations and deliverables and tasks from the team members.
- Manage workflow and approvals of shared document and completed tasks over the course of a project.
- Email alerts and other notifications whenever project deadlines are at risk.
- Integrated learning content and guides to help all team members communicate more effectively.

Alignment and Validation of Project Financials. In Six Sigma, the cost of poor quality (COPQ) and the program results must all boil down to dollars that are quantified. This is how Jack Welch and Larry Bossidy brought Motorola's Six Sigma from a initiative focused on productivity to an initiative focused on financial results.

Given this, any Six Sigma tracking infrastructure must provide a rich and fully configurable financial tracking and measurement capability for every project and for the overall project portfolio. Above and beyond a tracking and accounting capability, an enterprise system designed for Six Sigma initiative support should address sign-offs and validation of project-level performance and provide financial forecasts at each phase in the project. Features should include the following:

- Financial representative validation and sign off at each phase of the project.
- Reconciliation of the financial data from the projects with the financial data maintained within the organization's accounting systems.
- Spreadsheet file import and export between the tracking system, enterprise accounting systems, and the project-level tasks and tools.
- Flexibility to support any variety of financial processes, currencies, closing dates, and reporting formats required either at the project level or between the enterprise systems.

Allocation of Teams and Resources. At any level, process improvement is a team effort. Without the ability to classify and allocate resources, it is not possible to build and manage teams and improvements effectively across the initiative. Making sure that the most strategic projects have the right resources keeps the businesses very busy. The system's resource management capabilities must allow project leaders to allocate tasks and communicate with team members as needed to complete project deliverables.

More sophisticated enterprise solutions can track availability of resources through scheduling features and timesheets to track hours and costs both per project and over the entire initiative. Additional capabilities can effectively mirror the organizations HR files by not only tracking time and availability, but also skill levels, work history, and specific domain and project expertise. Systems should include the following:

- Ability to choose team members from lists matching both the desired Six Sigma training level required for the project as well as any project-related experience or affiliation with the business unity involved the project.
- Track time committed to each project and availability as needed for future projects.
- Align effectively with internal HR systems to both import and export data pertaining to each contributor's skill levels, compensation, and history of experiences.

Track Projects Individually or as a Portfolio. The success and importance of a project are usually measured in financial terms. Teams can only achieve these returns by keeping their focus on delivering on the individual set of metrics targeted for each project. For example, here are the bullets for a single project out of over 500 similar projects done for the year:

- Increased granular sulfate capacity by over 50 percent for $2 million additional income year to date:
  - Cause and effect analysis and FMEA applied to process.
  - Integrated Six Sigma, total quality, and maintenance excellence effort to exceed projected entitlement.
  - Developed and initiated new control tools and methodology.
  - Granular capacity increased with 570,000 tons produced (year to date) as compared to 449,000 tons for 1994 full year.

You can imagine trying to track all 500 projects to forecast the actual financial impact of the Six Sigma program. A key role in any enterprise management system is to not only track the tools and phases of any given project, but also the operations-level performance within the project to deliver targeted and predictable results. In Six Sigma, y = f(x, x, x). How well you manage your inputs (process variables) is how likely you will be to reach your key y's (project metric). Features include the following:

- Support for tracking multiple metrics within any given project.
- Support for any variety of reporting timeframes, tracking metrics, and targeted ranges.
- Capability to track differing project phases differently as the core processes targeted by the project are tuned and improved.

Tracking and Reporting of all Program Results. At the end of the day, the goal of any improvement initiative is to achieve results. Without some type of global view of the overall progress of each project and the progress of all of the projects as a portfolio, both the project teams, team leaders, and managing executives are blind to spotting trouble spots in the projects or recognizing areas of strong opportunity.

Dashboards and reports have long been a part of any enterprise-level management solution; however, with a Six Sigma tracking solution, these reports are tied directly to the project team and workflow to allow the managers to have immediate access to the project-level information and team members required to fix it. Reporting capabilities include the following:

- Drill down from high-level company-wide reports to project-level deliverables.
- Reliable and timely executive-level reporting data built from the ground up by first-hand project data.
- The flexibility to modify time, teams, areas of focus, and other variables to allow team to process and present data in whatever manner best suits their need.
- Capability to segregate reports and results along strategic lines or by specific project teams, geographies, or business units.
- Ability to roll up non-similar initiatives and differing team reports into unified executive reports to enable a "big picture" view for executives.
- Automation of ongoing project status reports to eliminate time-consuming data gathering and report creation that hampers many project teams.

Extensible Knowledge Sharing and Best Practices. Dramatic results are directly correlated to successful application of effective methodologies. Because Six Sigma programs consistently produce innovative ways to define business processes, the challenge is transporting innovations throughout the enterprise. Workflow and knowledge management capabilities work in concert within an enterprise system to clearly lay out defined best practices and available tools allocated for each project team. Even best practices can be continuously improved.

By maintaining a central knowledge base within the organization, successful projects can be easily leveraged and reapplied in other project teams with far greater speed and certainty than if the program needed to be developed independently. Features should include the following:

- Shared knowledge base of available project tools and reference documents.
- Repository of all completed project tools, documents, contributors, timelines, and results.
- Contextual recommendations and learning materials to provide guidance as to best practices tied to key process within the project workflow.

Adapt to Unique Business Needs and Systems. No one business is exactly like another and neither are their processes. This presents an ongoing challenge to software solution providers who must develop features that appeal to the wide cross-section of customer needs but be easily adaptable to the unique need of each project team. From seamless integration to the installed enterprise process, to financial and measurement systems, to the ability for each team to tune processes that work best for their needs, flexibility is a mandate of any enterprise-tracking solution. Capabilities need to include the following:

- Services-based architecture where differing contributions from differing applications can be widely shared and easily connect over standard Internet-configured networks.
- Configuration consoles where individual users and authorized team leaders can set personal settings, team settings, and rights and responsibilities for the entire team from a simple interface.
- Modular workflow capability where team and initiative leaders can adapt the system to reflective differing process improvement methodologies, processes unique to their organizations, or flexibility required for differences within specific business units.
- Import and export capability to freely move data and tools from any variety of desktop applications into a central knowledge base, project files, or financial segments of the system.

Results and Expectations. When teams and organizations take the step to begin managing their projects and performance management in an automated environment, it is common to see average project cycle times reduced by 50 percent or more and improvement of project performance by as much as 20 to 30 percent. The combined effect of simplified collaboration with the capability for more rapid awareness and response to problems in a project, enabled by the system, are the primary drivers behind the reduction in project cycle times.

The capability of the systems to allow the project teams to better evaluate ideas based on their strategic business needs and execute best practices are the primary drivers for improved average project performance. With a project portfolio management system in place, successful projects are more easily identified and more widely reused by teams anywhere in the organization. This capacity for leverage works to both lower project cycle times and increase average project performance equally.

ENTERPRISE SOLUTION PACKAGES (ALPHABETICALLY LISTED)

Instantis provides Enterprise Performance Improvement software for Global 2000 companies like Lockheed Martin, McKesson, and Xerox that have deployed Six Sigma and other structured, project portfolio-based business improvement methodologies. Instantis software automates the end-to-end execution, management, and reporting of these methodologies. With a unique capability to provide a bridge between strategic priorities and execution, Instantis solutions allow industry leaders to deliver improved financial results and better alignment of goals and activities throughout the organization. Learn more at www.instantis.com.

i-solutions helps organizations get more from their investment in critical business programs such as Six Sigma, IT, and R&D that drive growth and create value. They provide a range of enterprise software solutions, based on their flagship i-nexus product, that have been designed from the ground up to meet the real needs of Six Sigma VPs, CIOs, CTOs, and other business leaders. Combining the latest web technology with powerful ROI-driven functionality, these solutions are helping a growing number of global organizations to align their project portfolios with business objectives, accelerate the delivery of tangible benefits, and leverage what they learn in project execution. To find out more about how i-solutions can help you to reduce project cycle times, increase project success rates, and cut the cost of managing your portfolio, visit www.i-solutionsglobal.com.

PowerSteering Software provides executives with real-time visibility, strategy alignment, and enhanced program and team effectiveness. For more information, visit www.powersteeringsoftware.com.

DESKTOP SOLUTIONS

At the desktop level, today's solutions allow Black Belts and Green Belts to work from preconfigured templates and feature sets to support any variety of Six Sigma tools. Process mapping and project planning have all been fully automated and, in many instances, specifically refined for use in the context of any variety of Six Sigma projects. With these advanced tools, Black Belts and Green Belts can apply tools to more projects more easily and contribute results more quickly and widely than ever possible before. The essential Six Sigma desktop toolset requires a mix of the following capabilities:

- Statistical analysis
- Simulation
- Process mapping
- Project planning
- Reporting and presentation
- Spreadsheet

From no-cost open source solution suites to combinations of market-leading application suites costing as much as thousands of dollars per desktop, there is a wide selection of options and tradeoffs available to meet the project-level needs of the individual Black Belt and Green Belt. Some options to consider include the following:

- **Statistical Applications.** Minitab and JMP are the industry leaders in providing practitioners with the advanced statistical analysis capabilities to perform the correlations and regression analysis required to isolate root causes and prioritize key areas for improvement. For more information, visit either www.minitab.com or www.jmp.com.
- **Simulation Applications.** Crystal Ball is built upon a Microsoft Excel platform. Crystal Ball offers a package specialized for conducting simulations and determining distributions and optimizations that provide the foundation for more in-depth statistical analysis and process validation. For more information, go to http://www.decisioneering.com.
- **Microsoft Office (Excel, Project, Visio, and ppt).** For a one-stop desktop solution, the Microsoft suite is hard to beat for not only its scope of features, but also for its widespread acceptance and familiarity to business users around the world that serves to facilitate file sharing and ease of adoption. New developments from Microsoft have included templates designed specifically for Six Sigma and have begun to address capabilities for easier files sharing or integration into underlying enterprise applications. By virtue of its sheer size of its user base and feature set, at the desktop level, Microsoft is the ten-to-one leader in the tools arena.
  - **Excel**—The industry standard for complete spreadsheet, analytical, and financial content. Its veritable ubiquity and flexibility allow it to not only be close to a one-size-fits-all solution for most practitioner tools, but it also establishes the .xls file format as a standard for import and export of any numbers-based information between systems and individuals.
  - **Project**—Able to set schedules, manage dependencies, and assign resources. Project is easily the "next best" Six Sigma solution within the Microsoft suite. The application's rich feature set and ease of use make it the preemptive choice for project-by-project process management.
  - **Visio**—Though limited in its overall functionality relative to other MS applications, and despite the capability to mimic its functionality in other applications like Excel and PowerPoint, Visio is still the preferred choice for any practitioners who develop process maps on a regular basis.
  - **PowerPoint**—Like Excel is to spreadsheets, PowerPoint is also an industry standard for presentations and information sharing between teams. Though virtually unused as a Six Sigma project solution application, PowerPoint is essential for tollgate review presentations and reporting project and program results up and down an organization.

Other Applications. QI Macros, built upon an MS Excel platform, offers a comprehensive library of prebuilt Six Sigma tool templates to immediately provide resources for the project teams. For more information, go to www.qimacros.com.

iGrafx provides a family of business process analysis tools that help organizations understand, analyze, and optimize their processes for key corporate initiatives, including Six Sigma, Lean, and others. For more information, go to www.igrafx.com.

SigmaFlow is designed as a full-service desktop suite, and is the leading provider of Business Process Analysis (BPA) software that encompasses process design and simulation capabilities in a single, integrated environment. Process improvement professionals choose SigmaFlow because of the Six Sigma workbench capabilities—it saves time and dramatically improves results.

PLANNING YOUR SOLUTIONS

Like any Six Sigma project, implementing the software required to support your initiative requires a collaborative and coordinated effort to be effective. Well-thought-out and well-designed processes and planning lead to effective software solutions that can better automate and accelerate the performance of the team. The importance of this planning and preparation in the design of your supporting software infrastructure cannot be overstated:

- Map the process and needs of your teams.
- Strike the right balance of desktop and enterprise solutions.
- Leverage the processes and solutions you already have as much as you can.
- Invest in the solution most appropriate to your needs both for today and the future.
- Integrate your software vendors, Six Sigma trainers, IT teams, and business groups into the decision making.

EXECUTING SOLUTIONS

A common phrase used often in new initiative rollouts and change management is, "Start small (either in scope or feature), think big, and succeed often." Rome was not built in a day, and neither are most enterprise performance improvement initiatives.

To avoid organizational whiplash (and its correlated backlash), do not try to do too much too soon. It is very important that both from a process and technology point of view that you allow for the amount of time that it will take to get a system and its teams to full capability. It also takes some actual experience to understand what the final system should look like.

A suggestion for effectively starting small is to test your systems' processes within one or two smaller business groups in order to iron out any quirks. You will also identify any needs unique to your organization to be addressed in a controlled and safe environment before the system can be rolled out overall. Most Six Sigma companies deployed the enterprise system after the program was initially deployed. At that time, they had a better understanding of what they needed.

An alternative approach to starting small is to release the management system widely, but with only a limited feature set. This allows you to effectively promote that this new system is for the entire organization and that it will be the "way we do business," but allow the users to start with a manageable set of new requirements to learn and that you can be sure are ready to run. From there, the system can then be expanded incrementally as organizational learning and understanding increases.

On the subject of thinking big, groups need to understand that any effort to build and design solutions to automate and improve a business process will expand into a de facto process audit. Experienced teams recognize this and allow for this review phase in their rollout schedules. They will then bring contributors into the systems planning discussion, not only from the project teams who will be working on the system, but also from the financial and manufacturing groups who will have processes touching the systems and will need to be accounted for in the design and rollout of the systems and its underlying processes, workflows, rights, and responsibilities.

As for succeeding often, if you work in manageable steps according to a well-thought-out overall plan, there will be near certainty that your incremental progress will be positive. However, in the event that the fundamentals of the plan need revision or another look, by working in small incremental steps, you will have the opportunity to make more effective adjustments as you go and take advantage of the learning possible in each preceding step before planning your next ones.

TUNING AND EXPANSION

Successful software initiatives replicate successful improvement initiatives. The software champions work to continuously improve and strive to delight their customers. A successful software system is never "done." It should always remain a dynamic and living effort supported by continuous feedback and suggestions by its users and feature improvements by the vendors.

Good systems are designed to adapt to this feedback, and good systems managers are able to listen and facilitate to this dialog with the users. A key aspect in managing this feedback is to allow the time for users to settle in and spend time on the system. This can allow the managers to get a better view of what are larger issues that can be addressed within the system to best meet the greatest percentage of the individual requests.

Just as Six Sigma is not an isolated process targeted only for manufacturing, your Six Sigma software tools and enterprise systems do not have to be limited to Six Sigma project work. Today's solutions can be expanded easily to support project management needs in IT teams, supply chains, sales and marketing, and many more.

A key benefit to expanding the application of your software resources is that it creates a larger user base to gain value and improve not only the overall ROI of your software investment, but also the base of resources available to support them. By deploying an enterprise system that works in supporting your Six Sigma efforts, the same system will need only minor modifications to apply to other, future change initiatives. A good enterprise system now leads to an accelerated and sharper set of change initiatives in the future.

15 Leading Six Sigma for the Long Term

With Dick Scott

We have been focused thus far with successfully launching a company-wide Six Sigma deployment within 90 days. Your Six Sigma program has now been successfully launched. You must now implement the systems and processes needed to effectively lead the deployment over the long term. This is a critical step toward ensuring that you achieve a culture change and that Six Sigma becomes how you do your work. Failing to implement these required processes and systems is a sure guarantee that Six Sigma will simply become another flavor of the month.

The purpose of this chapter is to identify and clarify a *leadership roadmap* for Six Sigma. The Six Sigma leadership roadmap will intensify and institutionalize your Six Sigma program and make Six Sigma a model initiative. The deployment roadmap, while paying attention to transforming your company, has been based on John Kotter's eight-stage process of leading change. We are now ready to work on Stage Seven (consolidating gains and producing more change) and Stage Eight (anchoring new approaches in the culture).

Kotter's Stages Seven and Eight

Consolidating Gains and Producing More Change. Pending a successful launch that produces dramatic early results, the next step is to grow the Six Sigma initiative with other focused efforts. For example, AlliedSignal focused on manufacturing operations during the first year and then added emphasis to growth by launching Design for Six Sigma (DFSS), and later started Transactional Six Sigma. Each new effort has the same accountability to the financial targets as the first project.

This stage evolves Six Sigma to include all parts of the company. Usually, the first evolution for U.S.-based companies is to launch Six Sigma in Europe and Asia. As Kotter quotes, "Culture changes only after you have successfully altered people's actions, after the new behavior produces some group benefit for a period of time, and after people see the connection between the new actions and the performance improvement."

People will see over and over again Black Belts accounting for $250,000 to $1,000,000 to the bottom line, project after project. They will either want to be on a project team or be trained as a Black Belt. Simply, as stock values continue to go up, change becomes easier.

This stage prepares the organization to launch future changes to match the company vision. By really nailing your Six Sigma initiative, you will have lived through a change roadmap that will work for any future change you want to drive, regardless of what the change addresses. Learning how to effectively launch a change and then institutionalize it is a new core competency for your company.

Anchoring New Approaches in the Culture. After the Six Sigma deployment is successful, you have to start thinking about what you need to do to make sure the company's doing Six Sigma 10 years from now. Even the Six Sigma pioneer, Motorola, which launched Six Sigma in 1978, found Six Sigma lacking in the late '90s. You must understand that the change initiative—in this case, Six Sigma—has to be reviewed at least every three years with respect to reenergizing the initiative.

Anchoring an initiative is directly related to the infrastructure and systems developed to support the initiative. Fortunately, the very nature of Six Sigma addresses developing streamlined processes and systems. So, the systems developed to support Six Sigma should be user friendly and work quite well. For example, one method used by Fred Poses in AlliedSignal's early deployment days was to include each division's Six Sigma project plans in the annual operating plan. This straightforward anchor ensured that every plant manager knew how important Six Sigma was to Fred. An example of this method is displayed in Figure 15.1.

Plant	Project	Effect this Year ($ Millions)
Plant A		
	ME Reliability	0.6
	125 Capacity Increase	0.7
Plant B		
	Optimize Operations	0.3
	Parallel Drip Stripping	1.1
	Recover 38% HF	0.6
Plant C		
	Implement APC	0.1
	Distillation R&D	1
	125/124 Ratio	1.1
	Distillation Bottleneck	1
	Mechanical Reliability Improvement	0.8

Figure 15.1 An example of manufacturing plants adding their major Six Sigma projects to their annual operating plan.

Six Sigma must have something going for it judging by the number of CEOs who left Six Sigma companies for a non-Six Sigma company and then introduced Six Sigma to that company. This list is long, which indicates that Six Sigma should not be difficult to anchor in the culture, but methods of anchoring Six Sigma will be considered in your deployment plan.

FIVE STEPS TO LEADING SIX SIGMA

Let's talk about the five steps to leading Six Sigma. The most common question that arises when I talk to companies that are pondering Six Sigma is, "What are the most common failure modes in deploying Six Sigma?" The answer is found in Kotter's eight stages. Missing one of those stages or doing one of those stages poorly is a sure recipe for failure.

The five-step process I will show you proactively addresses some of the common failure modes. This simple (elegant) leadership roadmap will ensure the long-term success of your Six Sigma program. All your managers should commit this roadmap to memory.

By following this roadmap, you will be assured that your Six Sigma program will be anchored in your culture. Six Sigma becomes a documented business process rather than an extensive training program. You will have to train people in your company to apply Six Sigma principles to serious projects, and you have to train your leaders to lead Six Sigma effectively. The five-step process consists of these steps, as seen in Figure 15.2:

1. Select the right projects.
2. Select and train the right people.
3. Plan and implement the Six Sigma projects.
4. Manage Six Sigma for excellence.
5. Sustain the performance gains.

The preceding five steps represent process steps. The process inputs capture the key inputs needed to define the business needs. Recalling Larry Bossidy's three-element business model (external realities, financial targets, and internal activities), the inputs refer to the business strategic plan (STRAP) and annual operating plan (AOP) to define clear operating objectives. Add financial targets and information about your customers and the market, and you have the start of a complete set of process inputs. Figure 15.2 provides a schematic view of the leadership roadmap, including its inputs, process steps, and output. We will now cover each process step in detail.

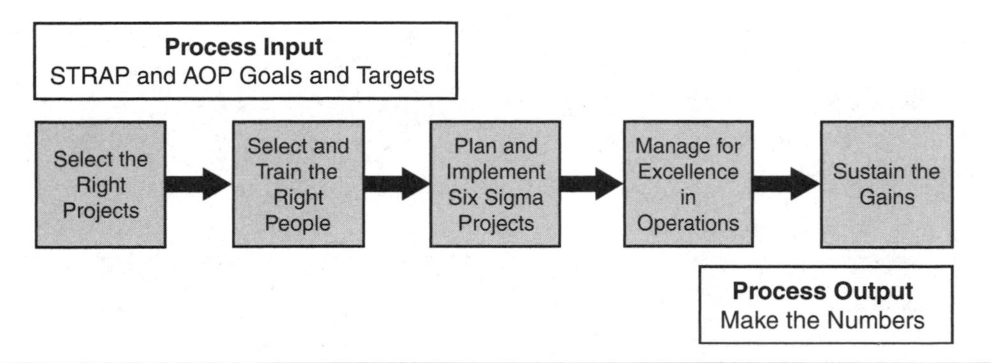

Figure 15.2 The Leading Six Sigma leadership roadmap—inputs and outputs.

Several of the key elements of this roadmap have been covered in detail in earlier chapters. Strategic process inputs, annual business goals, and targets are discussed in Chapter 2, "The True Nature of Six Sigma: The Business Model," and Chapter 5, "Strategy: The Alignment of External Realities, Setting Measurable Goals, and Internal Actions."

Step 1: Select the Right Projects is discussed in Chapter 9, "Committing to Project Selection, Prioritization, and Chartering."

Step 2: Selecting and training the right people is examined in Chapter 11, "Selecting and Training the Right People," and Chapter 8, "Defining the Six Sigma Infrastructure," covers the roles and responsibilities of those people.

Step 3: Plan and implement the Six Sigma projects is covered in Chapter 8 and Chapter 9.

Step 4: Manage Six Sigma for excellence is discussed in Chapters 5, 6, 8, 10, 12, 13, 14, and 16.

Step 5: Sustaining the performance gains is discussed in Chapter 6, "Defining the Six Sigma Program Expectations and Metrics."

Process outputs: make the numbers is also addressed in Chapter 6.

SIX SIGMA LEADERSHIP STEP 1—SELECT THE RIGHT PROJECTS

Selecting the right projects is the key step in institutionalizing Six Sigma. Meeting the process output of making the numbers is directly and strongly related to the aggressiveness and discipline of project selection. Performance breakthrough is tracked directly to the projects that produce the breakthrough. Chapter 9 gives you the details on identifying and prioritizing projects. For Six Sigma to be wildly successful, the deploying company must develop a clear process for accomplishing the selection of the right projects every year.

Selecting the right projects begins by making certain that everyone involved in Six Sigma understands the external realities, the company strategy, and the annual business goals and financial targets. Every project should have line of sight from each project to the corporate strategy or to the annual operating plan. Goal trees to the strategic and or annual business plan are a great visual depiction of the system (see Figure 15.3). Several manufacturing plants in AlliedSignal displayed these goal trees in their cafeteria to better communicate their Six Sigma activities.

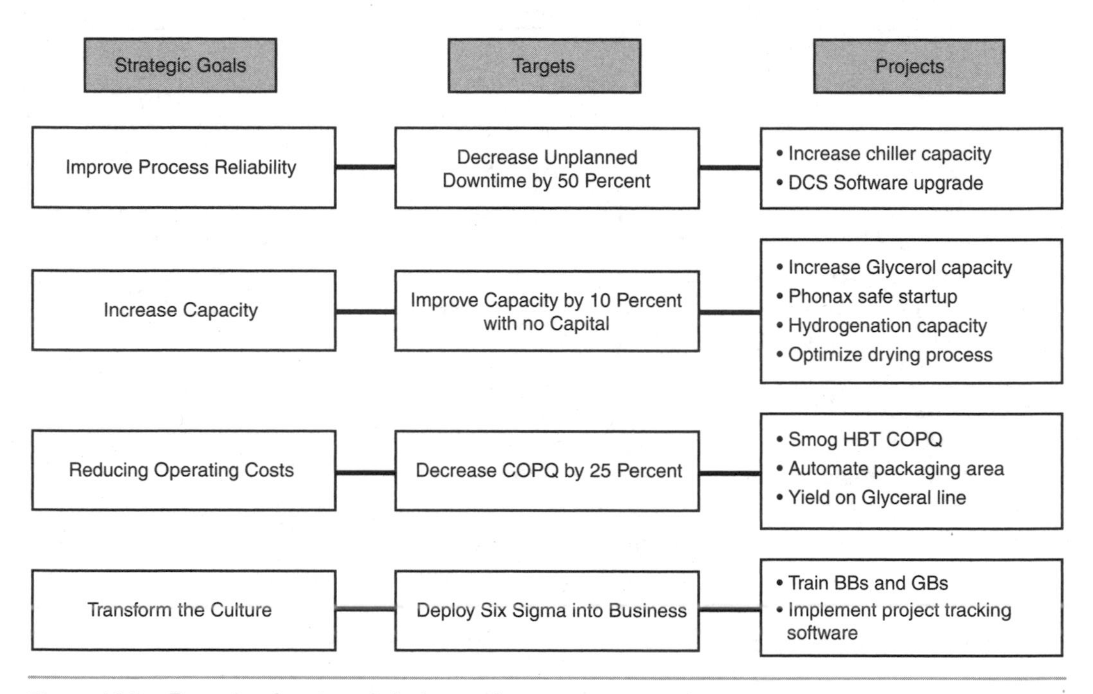

Figure 15.3 Example of projects linked to performance target and strategy.

This process step has a list of associated subprocess steps. They are

- **Process Inputs**: Strategic plan and AOP input from the business:
  1. Clarify the "big picture."
  2. Establish and document performance baseline and entitlement.
  3. Organize opportunities and prioritize.
  4. Select key projects with leadership buy-in.
  5. Check for accountability to business and personal performance objectives.
- **Process Outputs**: The output of the project selection step is a list of the top Six Sigma projects that will boost significant business productivity performance agreed on by the operations and business leadership and are linked to the strategy.

Objective of Step 1. The objective of selecting the right projects is to target the key business processes that are impeding significant productivity performance and are preventing the achievement of the company strategy. From his outstanding book, *The Power of Alignment*, George Labovitz (with Victor Rosansky) says that the greatest challenge that leaders have today is, "The *main thing* is to keep the main thing, the *main thing*!"

The diagram in Figure 15.4 shows the linkage between four components: our strategy, our customers, our people, and our processes. The processes link is what connects these linkages with Six Sigma. Our work is done through processes. If our processes are bad, our work is difficult and always lacking in some way. There are thousands of potential projects hiding within the processes in every business. The art of identifying the relatively few projects that will levy the most payback is the art of selecting projects.

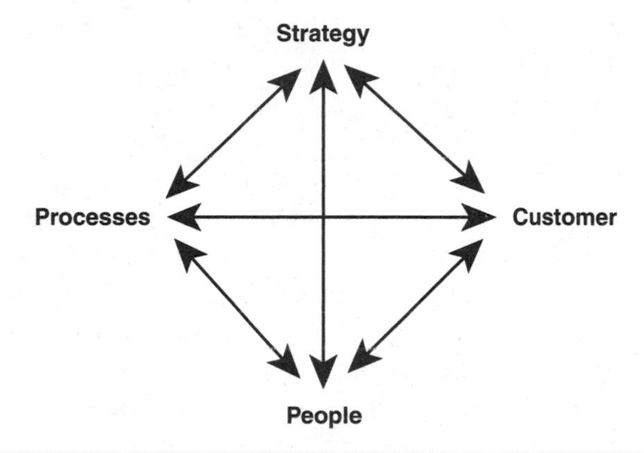

Figure 15.4 Example of linkages made in a Six Sigma deployment. Six Sigma is about making those connections.

Step 1.1—Clarify the Big Picture. Clarifying the big picture is the process of establishing linkages. Going to the business model, we must link the issues associated with our external reality to clearly stated financial and operational targets and coordinated internal activities.

The two substeps for Step 1.1 are to identify the business strategic objectives and to translate these into primary operational objectives for all levels and divisions of your organization. Along with these objectives, financial and performance targets will be defined. So if productivity improvement, cash flow improvement, and cost improvement are the important things, these must be made crystal clear to the company.

Any projects that don't address these three strategic areas are immediately called into question. Clarifying the big picture can be simple. Paul Norris, CEO of WR Grace, simply put one slide on the overhead projector: $56 Million. That's the level of pretax profit he was looking for at the end of the year. Everyone in the room was aligned with Paul. The level of urgency noticeably increased. Figure 15.5 shows a summary diagram of Step 1 of the leadership roadmap.

Step 1.2—Establish Organizational Performance Baseline and Entitlement. The next step is to have the organizations deploying Six Sigma establish their performance baselines and performance entitlements (the best possible performance given the current process). This step might be aimed at manufacturing only or manufacturing and other specified parts of the company.

One step might include doing an analysis of cost-of-poor-quality and waste in important processes. Conduct a "hidden factory" analysis to define manufacturing entitlement by identifying a comprehensive list of areas of waste or cost-of-poor-quality in the manufacturing process (i.e., brainstorm the question: "what are the sources of waste or inefficiencies in our manufacturing process?"). You can also do similar analyses for any process area—even legal, for example. Where is our legal department suboptimal? What is the financial benefit of fixing those things?

Group these lists of waste and COPQ into the broad categories of rolled throughput yield, capacity, productivity, COPQ, or product quality. Brainstorm a list of potential projects to contribute breakthrough wins in the businesses target areas of improvement (i.e., focus on the areas with the strongest links to the business needs).

Step 1.3—Organize Opportunities and Prioritize. Assign $ value to each project in the potential project list from baseline analysis. Plant finance must be involved in this step. Value should be stated in terms of $ value to operating income.

Next, estimate timing to complete (short-, medium-, or long-term) the project. Traditionally, we're looking for projects that take no more than six months to complete. We are ready to estimate resources needed to complete the projects (people, time, capital $, and expense $).

Finally, organize potential projects into a priority matrix (value, timing, and resources needed). Rank projects based on priority matrix output. Be sure to do a reality check on the top projects in the matrix.

Are these projects really doable? Is there a real link to the business (i.e., are the capacity improvement projects for products that are sold out?). Are the resource needs realistic? Then re-rank as needed. The output of this step is a list of the top *recommended* projects that will most significantly improve the results of the business. This process is described in detail in Chapter 9.

Step 1.4—Select the Key Projects with Leadership Buy-in. Review the recommended list of the top projects with the managers of Operations, Process Engineering, Sales and Marketing, Legal, HR, and so on. Within each function, agree on the top projects to target for this year.

Assign accountability to the owning manager. The owning manager reviews the projects with their business segment team to validate alignment with the needs, strategy, and financial targets of the business. Based on approval by the business segment team, the owning manager develops a project charter for each selected project with clearly defined objectives, timing, and linkage to the business. The output of this step is a complete project list, complete with project charters.

Step 1.5—Check for Accountability to the Business and Personal Performance Objectives. To ensure that the projects are linked to the future direction of the company, verify linkage for each project to the strategic plan. In addition, verify linkage/integration of each project with the AOP. Put them in the AOP if they are not already there.

Build the Six Sigma goal trees to demonstrate clear linkages. Include in the performance objectives of the owning manager on the proposed project list. The output of this step is clear business linkage and defined accountability.

Step 1 Outputs. The output of the project selection step is a list of the top Six Sigma projects that will yield significant business productivity performance agreed on by the business leadership. Figure 15.5 shows a summary diagram of Step 1 of the leadership roadmap with the five substeps.

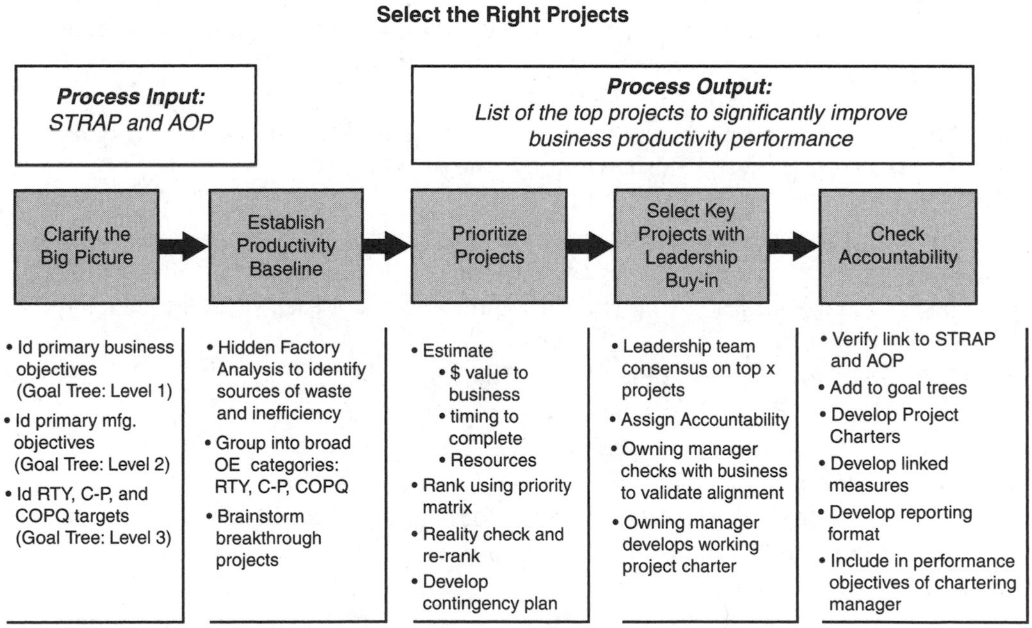

Select the Right Projects

Process Input:
STRAP and AOP

Process Output:
List of the top projects to significantly improve business productivity performance

Clarify the Big Picture → Establish Productivity Baseline → Prioritize Projects → Select Key Projects with Leadership Buy-in → Check Accountability

Clarify the Big Picture
- Id primary business objectives (Goal Tree: Level 1)
- Id primary mfg. objectives (Goal Tree: Level 2)
- Id RTY, C-P, and COPQ targets (Goal Tree: Level 3)

Establish Productivity Baseline
- Hidden Factory Analysis to identify sources of waste and inefficiency
- Group into broad OE categories: RTY, C-P, COPQ
- Brainstorm breakthrough projects

Prioritize Projects
- Estimate
  - $ value to business
  - timing to complete
  - Resources
- Rank using priority matrix
- Reality check and re-rank
- Develop contingency plan

Select Key Projects with Leadership Buy-in
- Leadership team consensus on top x projects
- Assign Accountability
- Owning manager checks with business to validate alignment
- Owning manager develops working project charter

Check Accountability
- Verify link to STRAP and AOP
- Add to goal trees
- Develop Project Charters
- Develop linked measures
- Develop reporting format
- Include in performance objectives of chartering manager

Figure 15.5 Summary of Step 1 of the Six Sigma leadership roadmap.

SIX SIGMA LEADERSHIP STEP 2—SELECT AND TRAIN THE RIGHT PEOPLE

Objective of Step 2. Step 2 showcases the process of identifying and defining roles of the critical people needed for successful completion of the key targeted projects, which were output from Step 1. Chapters 8 and 11 address this step in detail.

Step 2 Process Input. The input of this process step is a list of the top Six Sigma projects leading significant business productivity performance as agreed on by the division and business leadership.

Step 2 Process Steps. Process Step 2 consists of five subprocess steps. This step addresses selecting, systematically training, and providing time and resources for the students to complete their Six Sigma project work.

1. Ensure the right leadership and ownership.
2. Develop a training plan for the right people.
3. Provide the right training.
4. Dedicate time for the trainees to complete projects.
5. Ensure the right resources are available.

Step 2.1—Ensure the Right Leadership and Ownership. The manager or leader that owns projects visibly champions key projects by face-to-face education of their organizations on the goal trees. Clearly defining the linkage of each project to the strategy of the annual operating plan, and to beating the financial targets, develops a sense of urgency within the Six Sigma project teams to complete the projects.

Use these meetings to establish the WIIFM (What's in it for me?) for the owning organization for the key projects. The output of this step is clear communication of the business and manufacturing goal trees down to the key project level. The other output is to demonstrate that your leadership is counting the success of each project to reach its strategic vision.

Step 2.2—Develop a Training Plan for the Right People. The second step for Step 2 is to first identify potential Six Sigma candidates from across the company. You will be searching for candidates within the executive leadership, the Six Sigma Initiative Champion, Six Sigma Deployment Champions, Project Champions, Black Belts, Green Belts, and Yellow Belts. You will also be looking for candidates for Master Black Belts. This is a pretty complex group to train, and an annual plan is the key to making the training happen.

Now that you have a list of prioritized projects from Step 1, tentatively match these candidates to the Six Sigma project with the best fit. You may use the following suggested criteria to test the realistic fit of the candidate to the target project. The candidate:

1. Has an understanding of the targeted business process?
2. Has general technical acumen?
3. Has ability to work within and lead a team?
4. Has a bias for action?
5. Is willing to change the status quo?
6. Has a desire to work on the project?
7. Will be in his/her position long enough to finish the project?
8. Can be relieved of current duties for time needed for successful completion of the target project?
    a. For Black Belts, this must be >75 percent (100 percent for top five projects).
    b. For Green Belts, this must be 25 to 50 percent.
    c. For Yellow Belts, this must be >25 percent.

The final task is to document the training plan for the identified Six Sigma training candidates. This plan should be reflected in each segment of the company's annual operating plan. The output of this process step is a documented training plan identifying the Leaders, Black Belts, Green Belts, Yellow Belts, and Master Black Belts and their assigned projects.

Step 2.3—Provide the Right Training. A guiding coalition of Six Sigma function leaders and Six Sigma Champions will be responsible for executing the annual training plan. Course schedules and the training resources will be ready. Six Sigma training plans are extensive and include more training time than usual. For example, here are some estimates of Six Sigma training time for various programs:

1. Champion Training—three to four days
2. Executive Training—two to three days
3. Black Belt Training—four to five weeks
4. Green Belt Training—two weeks
5. Yellow Belt Training—four days
6. Master Black Belt Training—five to eight weeks

Figure 15.6 presents a straw man annual Six Sigma training plan. The word "wave" is synonymous with "class." The plan should include a high-level schedule such as this, as well as an estimate of the number of people to be trained in each event. You should be able to track the number of people trained in Six Sigma over time.

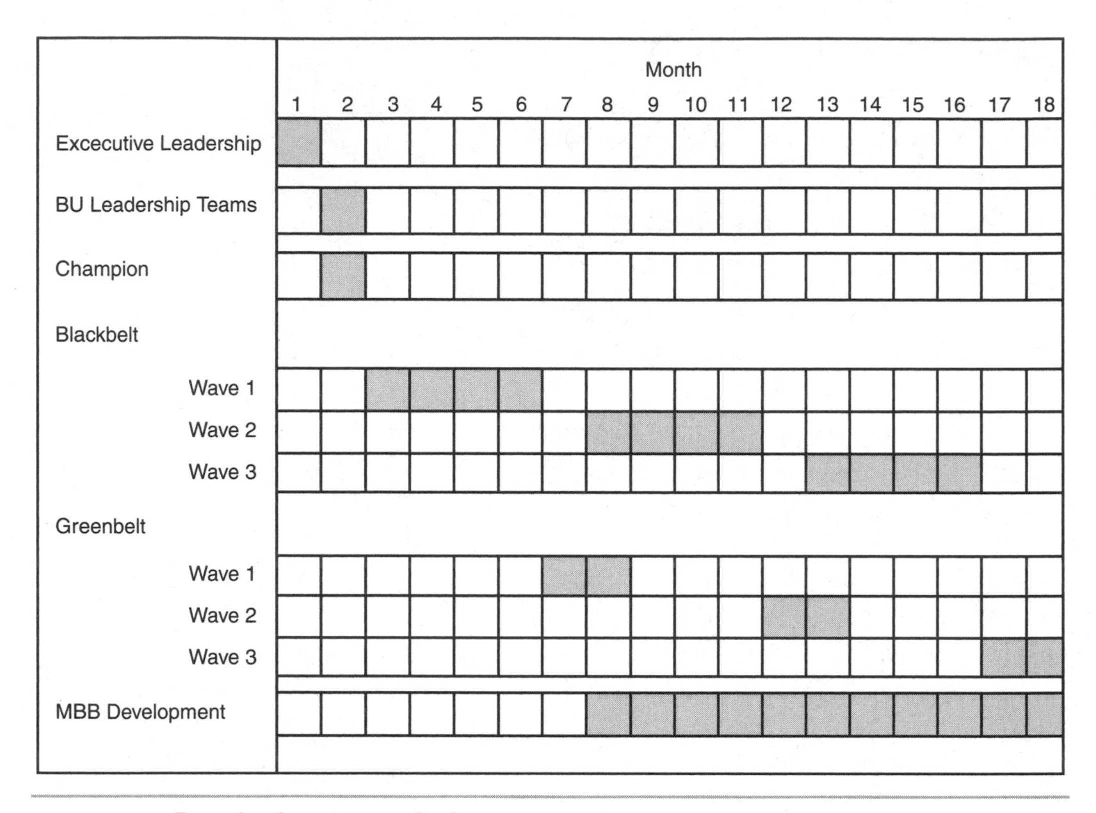

Figure 15.6 Example of a straw man Six Sigma training plan over the course of one year. The term "wave" is synonymous with "class." This plan is coordinated with project selection.

Table 15.1 demonstrates an actual two-year draft of a large company's Six Sigma training plan. Notice the number of Green Belt events on the plan. Once Six Sigma is deployed, this company reinvigorated Six Sigma by launching new programs with different focuses and different goals.

Table 15.1 Example of a Two-Year Six Sigma Training Plan

- Black Belts: 150
- DFSS Black Belts: 50
    - DFSS Master Pilot: Q1/96
- Process Improvement Leaders: 500
- DFSS Green Belts: 60
    - Piloted: Q3/95
    - Launch: Q1/96, Desplains
- Yellow Belt Pilot: 11/95
    - Roll Out: Q1/96
    - 500 Trained in 1996
- Master Black Belt Development Program: 4–6 Certified, Q4/96
    - Defined: 11/95
    - MBB Candidates Identified: (2/96)
    - Implemented Beginning: Q1/96
- Leadership One-Day Six Sigma Sessions: 20 Leadership Teams

Table 15.2 shows an actual three-year summary of Six Sigma training from a $14 billion sector of a large Fortune 500 company. This chart shows the commitment of that sector's leadership to ensure that Six Sigma became part of the company's culture. This program resulted in millions of dollars in financial benefit.

Table 15.2 Example of a Three-Year Summary of Six Sigma Training Activities

Training Metrics	1995	1996	1997
Master Black Belts	6	14	20
Black Belts	170	308	450
Product Development Black Belts	0	65	165
Green Belts	222	522	822
DFSS Green Belts	20	75	275
Yellow Belts	0	125	1,000
Total	418	1,109	2,732

Step 2.4—Dedicate Time for the Black Belts and Green Belts. A controversial issue that inevitably rears its ugly head during every Six Sigma deployment is the time the Belts should dedicate to their projects. This issue is covered in Chapter 13, "Creating the Human Resources Alignment." The suggestions range from 100 percent of their time to some percentage of their time. Our data clearly shows that the more time the Belts spend on their projects, the faster the projects get completed. If you have identified a project worth a million dollars, why would you want someone to work part-time on it? I'll present some recommendations.

For the top-five productivity projects in each functional area, I suggest you relieve Black Belts from their regular duties. Assign them to work 100 percent of their time for the 4–6 months they will spend on their key project. During this period, consider assigning them to report to the Six Sigma Project Champion (at least on a dotted-line relationship). Note: The people in this core project group should rotate out every year or so. Assign remaining Black Belts to work 75 to 100 percent of their time on their assigned projects.

Assign Green Belts to work >50 percent of their time on assigned projects. For each of the Black Belts and Green Belts, clearly define what work they will stop doing and who, if anyone, will pick it up so the needed time is really available.

The site Master Black Belt and finance person should provide continuous technical support to all Black Belts and Green Belts. Clearly set performance expectations for Black Belts and Green Belts by incorporating their project goals in their annual performance objectives. Make sure they understand and accept the changes needed in their work-style (shift from fire-fighting to continuous process improvement focus). Manage Black Belt and Green Belt performance actively. Assess regularly and support/discipline/coach as needed.

Step 2.5—Ensure the Right Resources Are Available. Just assigning a project to someone without dedicating the resources can be classified as cruel and unusual punishment. During the project selection process, leadership also is aware of required resources and commits the resources required for each project. Table 15.3 shows "mini-charters" for a small set of projects that indicates the leadership group was aware of who the resources were before the project was officially chartered. This table also shows cost estimates and capital estimates for each project along with an estimate of business benefits.

Table 15.3 Example of Actual "Mini-Charter" Detailing the Initial Requirement for Resource (Both People and Capital)

Project	Priority Score	Business Impact Financial ($$)	Business Impact Nonfinancial	Measurement	Resources/ People	Resource Capital 1998–1999	Resource Expense
Glycerol Capacity	189	$0.8M	$70 million sales to company	Pounds per day	Shah, Fuller, Ruelens, Mehta, and Weidenman	Unknown $100,000	Demos:
Automate Packaging Area	145	$1.9M	Revenue gains due to capacity improvement, depending on new market availability	Units shipped per day	Halll, Haynes, Shurnoski, and Buzzo	$600K	$100K
DCS Hardware/ Software	133	$3.0M	$5MM in sales	Pounds per day	Perkins, Deforio, Delaney, and Erpst	$50K	$25K
Training Plan for BBs/GBs	129	$8M	Improvements in cost, quality, and cycle time	Execute plan to 90 percent	Grubb, Facile, and Crown	N/A	$300K

So, identify the people with the skills and knowledge needed to help solve the problem and select those who can dedicate the needed time to contribute.

Assemble Six Sigma teams around the specific projects. Each team should be led or championed by a Black Belt or a Green Belt.

Step 2 Output: The right project owners and leaders with a shared commitment.

SIX SIGMA LEADERSHIP STEP 3—DEVELOP AND IMPLEMENT IMPROVEMENT PLANS FOR KEY SIX SIGMA PROJECTS

This is the step in the leadership roadmap where the rubber meets the road. Six Sigma is based on process improvement roadmaps. There are different roadmaps for different functions. There is a specific roadmap for manufacturing operations. There is another roadmap for transactional functions such as HR, legal, and purchasing. R&D and product development have their own set of roadmaps. There is even a version of Six Sigma that integrates Lean manufacturing into the roadmap.

Objective of Step 3: To translate the acquired process improvement roadmaps and tools into progress for each of the key targeted projects.

Step 3 Process Input: The right project owners and leaders with a shared commitment.

I will describe the manufacturing operations roadmap to give you an idea of the structure of process improvement. This roadmap is based on the Measure, Analyze, Improve, and Control (MAIC) process. I've shortened the roadmap in lieu of time and space for this section.

Step 3.1—Measure the Targeted Process.

1. Define and document key product, customer requirements (Key Process Output Variables), program objective(s), performance variables, and process specifications resource requirements. This analysis should be based on the evaluation of rolled throughput yield (RTY), cost of poor quality (COPQ), and capacity-productivity (C-P).
2. Create a detailed process map and define Key Process Output Variables (KPOVs) and Key Process Input Variables (KPIVs) for each step of the process. Create a cause and effect (QFD) matrix relating Key Process Input Variables to Key Process Output Variables. Prioritize the KPIVs. Perform an initial assessment of the process control plan.
3. Identify gauge capability requirements for each key output and input variable and complete gauge studies as required.
4. Perform a short-term capability study on KPOVs to establish "first look" capability and establish process baseline.

Step 3.2—Analyze the Targeted Process.

1. Complete a Failure Modes and Effects Analysis (FMEA) to determine input variables that have high risks.
2. Perform multi-vari studies to understand how Key Process Input Variables (KPIVs) vary in the factory and to get a first quantitative look at the relationship between KPIVs and KPOVs.
3. From the FMEA and multi-vari studies, prioritize a list of KPIVs to test impact on KPOVs.
4. Review program and develop roadmap for establishing characterization, optimization, and reliability. Focus on capability study results, FMEA, and other sources of engineering inputs.
5. Plan first steps for the Improvement step.

Step 3.3—Improve the Targeted Process.

1. Use design of experiments (DOEs) to characterize the influence of the KPIVs on the KPOVs and to define sensitivity of KPOVs to changes in KPIVs. Identify and verify *critical* KPIVs using sequential DOEs and Evolutionary Operations (EVOPs).
2. Use DOE methods to establish optimal operating windows for the KPIVs.

Step 3.4—Control the Targeted Process.

1. Develop, document, and implement process support systems (training modules, maintenance plans, troubleshooting guides, process procedures, etc.) for a long-term capability run.
2. Demonstrate validity of operating windows through a long-term capability study.
3. Conduct final review of the program, write final report, and update support system documentation.

SIX SIGMA LEADERSHIP STEP 4—MANAGE SIX SIGMA FOR EXCELLENCE

The objective of Step 4 in the Six Sigma leadership roadmap is to provide the necessary on-going leadership and management support and guidance to ensure optimum results of the Six Sigma process in your organization. This step consists of six process substeps.

The lucky thing is these are the same steps for driving any initiative. The six substeps are as follows:

1. Stay focused.
2. Actively champion resources needed for progress.
3. Frequently review progress of projects.
4. Reality check the real business impact.
5. Continuously communicate progress.
6. Reward, recognize, and discipline performance.

Now, let's review each of the six substeps. Figure 15.7 shows a graphic of Step 4.

Step 4.1—Stay Focused. Common to many initiatives that work, the ongoing support and pressure from the organizational leaders are necessary. Success in this step is the function of clear accountability for results at all levels of the organization. This accountability leads to distinct behavior changes.

Success is related to maintaining focus on the goals of the projects and the long-term direction of the initiative. Black Belts keep projects focused on the key success factors for the business. Regular reviews of breakthrough projects provide leadership to the Black Belts. Review of the project's benefits and contribution to AOP and strategic plan goals is essential.

For example, the project-owning manager (Champions) will make sure their people are sticking to the project charter and progressing to the committed timeline. They will also actively manage Black Belt and Green Belt project focus and organizational leadership, and do a "direction check" quarterly.

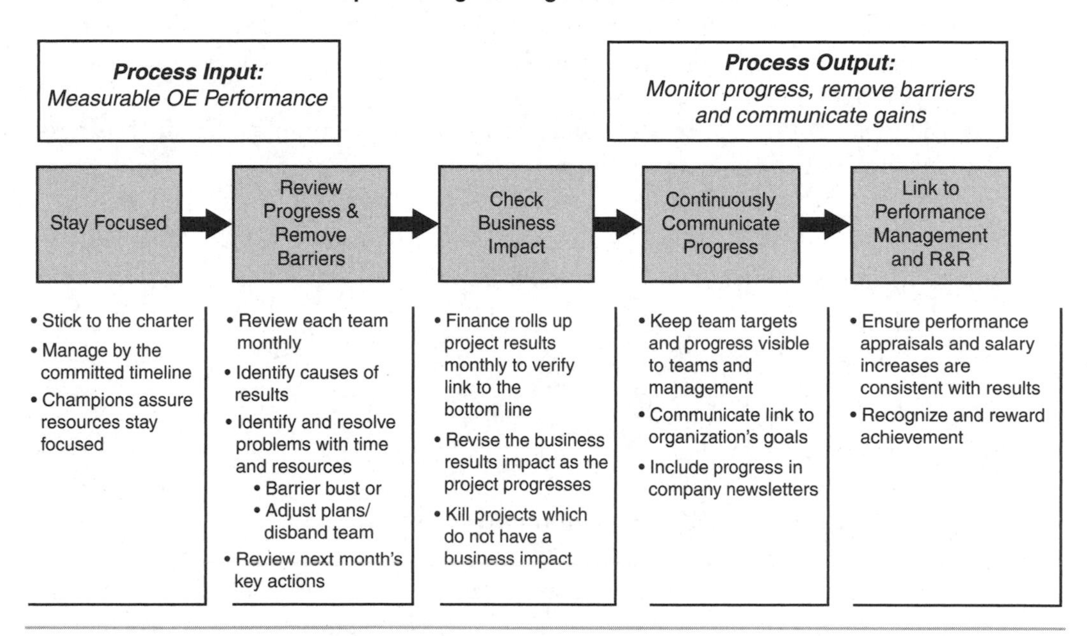

Figure 15.7 Step 4 of the Six Sigma leadership roadmap with substeps.

Step 4.2—Actively Champion Resources Needed for Progress. The fastest way to undermine a Six Sigma deployment is to let people block the Six Sigma teams from getting the resources they need. That's why it's so important for the organizational leaders to do a frequent reality check on resources. Is this a barrier for progress?

It's also essential to remove barriers to success and to barrier-bust where needed to get resources for key projects. If a team cannot get investment needed for progress, declare defeat and quickly move the Black Belt or Green Belt to another project. *Note: This should not happen if good project selection was done.*

Step 4.3—Frequently Review Progress of Projects. The easiest way to institutionalize Six Sigma is to review the program frequently at all levels. The plant manager may review his/her program once per month, the business president may review the program for his/her business once per quarter, and the CEO would review once per quarter as well.

For example, a manufacturing plant would plan to report monthly Six Sigma progress to their business leadership (in a flash report format). Black Belts hold monthly Six Sigma progress reviews with the group of project owners or Champions. The owning managers would schedule one-on-one reviews with Black Belts and Green Belts at key milestone dates to recognize progress, identify barriers, and reinforce expectations.

Champions summarize monthly the Six Sigma results of the key projects. The objective of these reviews is to sustain the drive for implementation and closure! *Note: I suggest having the owning managers occasionally report on the progress of the projects they own instead of Black Belts.*

Quarterly reviews of the entire Six Sigma system should be conducted at the business level. The focus is on the following:

- Project identification
- Project tracking
- Communications
- Accuracy of financial impact assessment
- Time commitment of Black Belts
- Project review processes

Findings from the review should be used to improve system weaknesses and promote strengths.

Step 4.4—Reality Check the "Real" Business Impact. To ensure that Six Sigma does not turn into a TQM program (lots of action and no results), it is crucial to have finance groups roll up project results monthly and verify link to "bottom-line" results (i.e., operating income). This action also gives the project champions an opportunity to revise the business results impact if needed as the project progresses. If necessary, the organization must learn that it's okay to "kill" projects that are failing the business results criteria. This decision should be in the hands of the owning manager, finance, and the business leader.

Benefits from each project are tracked and validated by the financial community to determine the project impact to the bottom line. Finance is involved in project chartering as well as project reviews to ensure

- Financial goals are properly calculated.
- Financial results are bridged to the bottom line.

Business Managers are required to sign off on projects in support of their business impact. Therefore, Business Managers must be included in project reviews when financial and business impacts are being reviewed.

Step 4.5—Continuously Communicate Progress. As with any change program, communicating program success constantly is so important. This communication might be monthly for a function as they progress toward their goals on using two-way communication. Also important is linking to the organization's performance in Six Sigma to various company award and recognition meetings. The Six Sigma communications plan is the foundation upon which the Step 5 activities are based. Constant communication is the hallmark for success.

Step 4.6—Reward, Recognize, and Discipline Performance. To anchor Six Sigma into your company, you must make it personal. Rewarding and recognizing fine performance is necessary, and driving discipline in performance is just as important. Ensure that Black Belt and Green Belt performance appraisal and salary increases are consistent with their annual performance objectives.

Reward and recognition programs must be appropriate for Black Belts/Green Belts and teams for significant results. Six Sigma teams will be performing at levels not previously seen in your company, which may prompt a revision of your current programs.

Everyone asks the question, "WIIFM?" Reward and recognition does not have to be monetary in nature, but reward systems should be consistent across the company, divisions, and plants. Rewards are correlated to the success of the project and the benefits to the business. Recognition should be public and formal at plant, division, and company levels. Ideas for rewards and recognition include the following:

- Team celebration day
- Gift certificates
- Trophies/plaques
- Company gifts
- Publication/invitation to present at company and corporate levels

Six Sigma Leadership Step 5—Sustain the Performance Gains

In any organization, the two most difficult things to accomplish are (1) implementing a new system and (2) getting rid of the old system. Your Six Sigma projects are focused on using fact-based roadmaps to reengineer a business process to enhance its performance.

Black Belts claim success when they can walk away from the process and the solutions continue to be effective. Your projects will result in developing a new system with which to perform better and perform consistently. Your organization has to protect the new system and prevent the organization from going back to the old system. There are four substeps to step 5:

1. Implement effective control plans.
2. Conduct regular training.

3. Review the Six Sigma system quarterly.
4. Continually identify key projects.

Step 5.1—Implement Effective Control Plans. Control plans are the instrument to prevent performance from slipping back to the old level. The new process must be thoroughly documented and all process participants trained on that documentation. This would include documenting control plans, and training both operators and supervisors on those control plans. The next step is to establish metrics and a tracking system to measure control plan effectiveness. Finally, the last step is establishing regular internal audits to assess the effectiveness of the control plan.

Step 5.2—Conduct Regular Training. Training at all levels is important in sustaining the gains. The types of training that occur frequently and systematically are as follows:

- Tools training
- Leadership training
- Team skills training
- Business training

This step entails the development of a rigorous training process. Training of new operators and re-emphasizing methods to existing operators sustains long-term performance. Process knowledge must be effectively transferred to operators who are empowered to take control of their process. Training of support personnel (maintenance technicians, QA/QC personnel, etc.) will also be developed and implemented for their specific support roles. The focus of the knowledge transfer changes to the key process requirements.

Step 5.3—Review Six Sigma System Quarterly. Creating a survey or formal assessment is the primary action in this step. One of the best examples was the Quality Systems Review (QSR) created by Motorola to assess the health of Six Sigma throughout the company. You will also need to audit project results.

So, you can use the Six Sigma process survey to identify any areas needing improvement. Once these areas have been identified, the next action is to implement needed changes and check for results.

Step 5.4—Continually Identify Key Projects. The foundation for identifying future projects is to maintain a common database to collect productivity or product quality improvement ideas. Rely on new ideas generated by the workforce. All your organizations must habitually maintain a running list of the top "next 10" key projects and assign Black Belts and Green Belts as they become available. This is sometimes referred to as the "project hopper." An easy question to ask a Champion is, "What are your next 10 projects?"

This last step of the leadership roadmap is important for truly anchoring Six Sigma into the culture. This is where the discipline of Six Sigma starts to happen consistently. To create a process change, the goal is to implement it quickly and have it last for years.

Six Sigma Handbooks and Other Anchors

During the front end of a Six Sigma deployment, have the Six Sigma Steering Team design a Six Sigma Handbook to document the way the initiative will be managed. The handbook is a great tool to use for company-wide training and for bringing new employees on board quickly. The following is a sample outline of a Six Sigma handbook.

Six Sigma Handbook

I. Cover Letter—Company CEO

II. Six Sigma Overview

III. Mission and Goals
- a. Metrics/ Entitlement
  - 1. Definitions and Calculations
- b. Key Reporting Mechanisms

IV. Strategy
- a. Strap/AOP
- b. Six Sigma Planning

V. Roles and Responsibilities
- a. Division Champion
- b. SBU Champion
- c. Site Champion
- d. Master Black Belt
- e. Process Improvement
  - 1. Black Belts
  - 2. Green Belts
  - 3. Yellow Belts

VI. Six Sigma Projects
- a. Method of Selection, Prioritization, and Assignment
- b. Chartering
- c. Tracking and Review System/Procedures
- d. Process for Closing Out Projects and Key Deliverables
- e. System for Archiving Projects

VII. Six Sigma Training
- a. Structure/Format

b. Prerequisites/Requirements
1. Calendar
c. Black Belt
d. Green Belt
e. Yellow Belt
f. Analytical/Lab Green Belt
g. Master Black Belt
h. Business Leadership

VIII. Certification Requirements
a. Black Belt, Green Belt, Master Black Belt, Champion
1. Results
2. Deliverables
3. Approval

IX. Six Sigma Tools and Deliverables
a. Roadmaps, Methodology, and Tools
b. Supporting Tools

X. Process Control System
a. Process Map
b. Correlation Matrix
c. Control Plan
d. FMEA

XI. Infrastructure Support/Data Systems
a. Minitab
b. Downtime Tracking
c. Project Tracking
d. Microsoft Project
e. KPV Control System

XII. Six Sigma Plant Diagnostic
a. Objective
b. Description
c. Process

Six Sigma Assessment. Another Six Sigma anchor is a Six Sigma assessment or diagnostic. Motorola created the Quality Systems Review to assess Six Sigma throughout the company. This was a company-wide and centrally controlled process that kept the focus on Six Sigma. The Six Sigma assessment can be a simple questionnaire or a more complex process that includes numerous site visits. An example of a survey questionnaire item follows. The example includes the survey item, which is scored on a 1 (Poor) to 4 (Excellent) scale; each item would have four descriptors, one for each number 1 through 4.

1. We select the right problems to work on:
   1.1 We randomly determine projects.
   1.2 We use baseline data to select projects.
   1.3 We use baseline data, and select projects by priority.
   1.4 We use baseline data, select by priority, and link projects to strategic plan.

This particular set of examples follows the Six Sigma leadership roadmap we have been discussing. By surveying your people and summarizing the scores, you will quickly discern where you are strong and where you are weak. Examples of Six Sigma survey items are as follows:

1. We select the right problems to work on.
2. We select the right people to be BBs/GBs.
3. We manage for the right performance.
4. Our Six Sigma goals are defined and understood by all employees.
5. The role of the Black Belts and Green Belts are well defined and understood by the organization.
6. Our overall Six Sigma process is effective.
7. We identify primary business objectives as the basis for Six Sigma projects.
8. We use COPQ, RTY, and CP to identify new projects.
9. Our Six Sigma projects are of a manageable size.
10. We organize project prioritization information into a priority matrix (value, timing, resources needed).
11. Projects are doable and linked to business needs (a reality analysis).
12. Projects are linked to the strategic plan, included in AOP and clearly defined on goal trees.
13. We use defined criteria for BB/GB selection and project assignment. BB/GB understands the plant process, technically strong, able to work within and lead a team process, have a bias for action, and are willing to change the status quo.
14. Our BBs/GBs have the right level of experience to be successful and to establish credibility with operators and supervisors.
15. Which statement best describes how the owning manager supports BB/GBs?
16. We identify the people with knowledge of the problem and the skills to solve the problem and give them dedicated time to work on the problem.

17. We stick to the project charter and manage by the committed timeline.
18. We frequently do a reality check on resourcing ($ and people).
19. We frequently review progress of projects.
20. We implement and close projects.
21. Finance audits projects and verifies project results. Projects that fail the business results test are stopped. The bottom-line analysis is used to track monthly progress.
22. BB/GB performance appraisals are based on their Six Sigma progress.
23. BBs/GBs receive appropriate R&R for significant results.
24. Projects have effective control plans that are documented, implemented, and periodically audited to ensure that we sustain the desired results.
25. BBs/GBs receive appropriate leadership training.
26. Our Six Sigma efforts are providing the expected $ benefit.
27. Plant management reviews the prioritized project list and results of the "reality" analysis.
28. The Business Area manager shares projects with business (segment, marketing, etc.) leaders to validate alignment with business objectives.
29. Projects have the right leadership and ownership.

More complex and yet complete assessments are available. For example, the Motorola QSR has 11 subsystems and generates scores for each and profiles for each subsystem. It takes one week to complete and provides business-wide information. So, by following the Six Sigma leadership roadmap as described in this chapter, you will be well underway to making Six Sigma one of the best initiative deployments your company has ever seen.

16 Reinvigorating Your Six Sigma Program

With Debby Sollenberger

Why in the world, you might ask, would you write a book on deploying Six Sigma in 90 days and end the book with a chapter about reinvigoration? Every company that has deployed Six Sigma has always set forth some actions to keep the program in tune to the company.

Any change initiative becomes different as the company learns the initiative and what it can do. For example, moving Six Sigma to parts of the company that were not initially heavily involved usually reinvigorates Six Sigma. An example of that would be a company that initially focuses on manufacturing improvement and then moves Six Sigma into product design and sales and marketing.

Also, the business reality for a company is not stagnant. As the business reality changes, Six Sigma must be modified to meet the new gaps in the market. I will be addressing three topics in this chapter:

- Critically analyzing your deployment.
- Back to the basics.
- Time to extend Six Sigma.

All of these topics apply to a new deployment as well as a deployment that is a year or two old. Going in to the first 90 days of a deployment is a good time to see potential failure modes and deployment opportunities. Potential reinvigoration targets are gaps in executing your strategy or business model. If you recall, Larry Bossidy's business model from his book, *Confronting Reality*, includes linking your external realities, financial targets, and internal activities. Here's how you can use those three components to look for reinvigorating opportunities:

- **External Realities**
  - Have you identified gaps in customer satisfaction?
  - Are there new markets to attack?
- **Financial Targets**
  - Is it time to set new financial targets or operating metrics?
  - Financial: operating margins, cash flow, revenue growth, ROE.
  - Operating metrics: productivity, quality, cost of poor quality.
- **Internal Activities**
  - Is your strategy linked to the annual operating plan?
  - Are your Six Sigma projects linked to the strategy and annual operating plan?

There are several reasons to always think about reinvigorating Six Sigma. Any change initiative has a risk of failure. Bossidy even said that if Six Sigma wasn't successful within AlliedSignal, he would never be able to successfully launch another change initiative. Because Six Sigma is complex, it is impossible to forecast all failure modes in advance and prevent them. The purpose of this book is to minimize the number of events that can go wrong. To make change work requires constant vigilance and correct innovation and action. By making sure Six Sigma works, you'll be known as a

- Company that executes.
- Company that deploys and achieves action.
- Company that has a core competency of deploying initiatives.
- Company that leverages human creativity and potential.

In addition, the final impacts of a successful Six Sigma program are the factors every company would like in place.

- Every employee understands the company's business, goals, and vision.
- Every employee knows how he or she contributes to the company.
- Every employee knows how to improve processes.
- Every employee knows how to solve problems.
- Every function works together seamlessly.

To assess your Six Sigma deployment, you can use Larry's advice on leading a change initiative:

- Learn the guts of it.
- Invest your time and energy.
- Pick the right people to initiate it.
- Be courageous.

The quality of deployment excellence is directly linked to how you and your fellow leaders accomplish the preceding. Change is not easy, but it is a lot of fun when it works.

CRITICAL ANALYSIS OF YOUR SIX SIGMA DEPLOYMENT

The first step in reinvigorating Six Sigma is to do a program assessment. In this assessment, you must consider the following deployment factors:

- **The Power of Assessment: Analyze**
  - Leadership and accountability
  - Program metrics
  - Deployment plan
  - Communication plan
  - People
  - Projects
  - Alignment
  - Speed
  - Results

By doing an honest assessment of those factors, you will always develop ways to make your Six Sigma program better. Gaps in any of the preceding will tend to slow the impact of Six Sigma in your company. A survey is a great way to get the assessment started. The following example survey has 28 questions, each scored on a 1 to 4 scale. Higher numbers on the scale represents better performance on that item. This survey focuses more on the day-to-day leadership of the program. These are good questions to answer when starting the deployment so you ensure that these items are linked to the deployment plan.

The focus of the Six Sigma assessment will identify the degree to which new Six Sigma behaviors are replacing the old behaviors. Your people should show a tendency to manage and constantly improve work processes. The old behaviors should start disappearing and the new Six Sigma behaviors are seen much more often.

SIX SIGMA SURVEY

1. We select the right projects to work on.
2. We select the right people to be Black Belts and Green Belts.
3. We manage for the right performance.
4. Our Six Sigma goals are defined and understood by all employees.
5. The role of Black Belts and Green Belts are well defined and understood by the organization.
6. Our overall Six Sigma process is effective.
7. We identify primary business objectives as the basis for Six Sigma projects.
8. We use COPQ, RTY, and CP to identify new projects.
9. Our Six Sigma projects are of a manageable size.
10. We organize potential projects into a priority matrix (value, timing, resources needed).
11. Projects are linked to STRAP, included in AOP, and clearly defined on goal trees.
12. Projects are doable and linked to the business needs (a reality analysis).
13. We use defined criteria for Belt selection and project assignments. Black Belts and Green Belts understand the plant process, are technically strong, are able to work within and lead a team process, have a bias for action, and are willing to change the status quo.
14. Our BBs/GBs have the right level of experience to be successful and to establish credibility with operators and supervisors.
15. We identify the people with knowledge of the problem and skills to solve the problem and give them dedicated time to work on the problem.
16. We stick to the project charter and manage by the committed timeline.
17. We frequently do a reality check on resources ($ and people).
18. We frequently review progress of projects.
19. We implement and close projects.

20. Finance audits projects and verifies projects results. Projects that fail the business results test are stopped. Finance tracks cumulative financial results to be reported monthly.
21. BB/GB performance appraisals are based on their Six Sigma progress.
22. BBs/GBs receive appropriate R&R for significant results.
23. Projects have effective control plans that are documented, implemented, and periodically audited to ensure that we sustain the desired results.
24. BBs/GBs receive appropriate leadership training.
25. Our Six Sigma efforts are providing the expected $ benefit.

ONLY MANAGERS/SUPERVISORS NEED TO ANSWER THE NEXT THREE ITEMS

1. Plant management reviews the prioritized project list and results of the "reality" analysis.
2. The Business Area manager shares projects with business leaders to validate alignment with business objectives.
3. Projects have the right leadership and ownership.

BACK TO THE BASICS: CRITICAL QUALITY DIMENSIONS

When looking at gaps in a Six Sigma deployment, it's always good to go to some critical dimensions to look for clues for mediocre performance. These dimensions would include the following:

- Leadership Commitment
- Leadership Involvement
- Strategic Alignment
- Project Focus, Identification, and Pipeline
- People
- Methodology
- Communication Plan
- Discipline
- Cultural Readiness
- Financial Results

- Project Tracking
- Partners

LEADING SIX SIGMA

We'll start with leading Six Sigma. Chapter 15, "Leading Six Sigma for the Long Term," covered the Six Sigma leadership roadmap. The roadmap looked like this:

- **Inputs**
  - Strategic and Annual Business Goals and Targets.
- **Step 1:** Select the Right Projects.
- **Step 2:** Select and Train the Right People.
- **Step 3:** Develop and Implement Improvement Plans.
- **Step 4:** Manage for Excellence.
- **Step 5:** Sustain the Gains.
- **Outputs**
  - Make the Numbers.

Chapter 15 covers each step in detail. We will look at the same steps and focus on potential failure modes. In a sense, this is another look at assessing your deployment but at a system level. Does your Six Sigma system look good and does it work?

Step 1: Select the Right Projects. Figure 16.1 shows the key actions associated with Step 1. This step ensures the link of strategy to the projects, delineates accountability, and establishes the current performance and productivity baselines. This step focuses heavily on prioritizing breakthrough projects and making sure training is included in the Annual Operating Plan. Right after Figure 16.1 is a list of the common failure modes for this step.

Select the Right Projects
• Clarify big picture using strategic plan • Establish plant/area productivity baseline • Prioritize projects based on value, resources, and required timing • Select key projects with leadership buy-in • Check accountability: business and personal

Figure 16.1 A summary of the actions associated with Step 1: Selecting the Right Projects.

FAILURE MODES FOR SELECTING THE RIGHT PROJECTS

Each of the five steps in leading Six Sigma have associated failure modes. If one understands what the failure modes could be ahead of time, preventing the failure modes from occurring is much easier. Four common failure modes for selecting the right projects would include the following:

- Poorly defined strategy.
- Projects not tied to financial results.
- Poorly defined project scope, metrics, and goals.
- Many projects lasting more than six months.

By assessing the Step 1 actions and the failure modes, ways to improve your ability to select the right projects will be improved. This project selection system will be usable within any follow-on improvement initiative.

Step 2: Select and Train the Right People. Figure 16.2 summarizes this step. Selecting and training the right people is the purpose of Step 2. This step addresses selecting the Six Sigma Champion, Master Black Belts, Black Belts, and Green Belts. The training plan is developed as part of this step, which includes leadership and Belt training. After Figure 16.2, the common failure modes for this step are listed.

Select and Train the Right People
• Ensure the right leadership and ownership • Develop a training plan • Dedicate time for training and application • Ensure the right support resources are available

Figure 16.2 A summary of the actions associated with Step 2: Selecting and Training the Right People.

FAILURE MODES FOR SELECTING AND TRAINING THE RIGHT PEOPLE

Four common failure modes for selecting and training the right people are included in the following bulleted list. There are, of course, many more failure modes you will be considering throughout your deployment.

- Wrong people assigned to the projects.
- Leadership doesn't understand their role.
- No centralized training plan.
- Lack of understanding of Six Sigma resulting in poor support from Finance, IT, HR, Maintenance, and QC Lab.

These failure modes are easily preventable by setting the right actions in the deployment plan.

Step 3: Develop and Implement the Improvement Plans. Figure 16.3 summarizes Step 3 of the leadership roadmap, developing and implementing the improvement plans. This is the muscle of Six Sigma leadership. The projects have been selected, validated, and chartered. The students have been trained in Six Sigma methods and the teams are in place. There are now dedicated Black Belts, Green Belts, and resources.

Support resources are at ready alert to provide support to the Six Sigma teams as needed. You even have a network of Master Black Belts floating around mentoring the Black Belts and teams. Your training has been customized to target groups and the projects cover a wide array of applications (Product Development, Administrative, Asset Dependability, Marketing, Sales, Human Resources, and Technology).

Develop and Implement Improvement Plans
• Measure Process • Analyze Process • Improve Process • Control Process

Figure 16.3 A summary of the actions associated with Step 3: Developing and Implementing Improvement Plans.

FAILURE MODES FOR DEVELOPING AND IMPLEMENTING IMPROVEMENT PLANS

The action of developing and implementing improvement plans represents the engine of Six Sigma to get things done and to attain the aggressive goals you have set. Six of the common failure modes include the following:

- Belts have little or no time to work on their projects.
- Large project teams and infrequent project meetings.
- Poor support from rest of the organization.
- Insufficient project support, and/or project support from a BB with limited experience.
- Lack of technicians and operators when you need them.
- Focus in on training, not improvement.

These failure modes directly relate to the sense of urgency the organizations within a company have about getting the results from the projects. The other aspect relates to how well the teams are lead, managed, and supported. Six Sigma is not a training game but a serious effort to improve the business. If the first step is done well, and great, believable projects are selected, the teams have a good sense of urgency around each project. Six Sigma projects and teams are integrated directly into the day-to-day operations of any organization.

Step 4: Manage Six Sigma for Excellence. Figure 16.4 shows Step 4, managing Six Sigma for excellence. Step 4 is the key to institutionalizing Six Sigma. The Champions reviewing their projects at least once per week and consistently reaffirming the potential financial impact of the projects from a finance point of view is critical. Each Champion should plan on spending at least 30 minutes per week per Belt. Effective communication of the program and a strong R&R scope will be the grease that keeps the wheels rolling.

Manage for Excellence
• Stay focused • Frequently review progress and remove barriers • Check real business impact [Finance integration] • Continuously communicate progress • Link to performance management and R&R

Figure 16.4 A summary of the actions associated with Step 4: Managing for Excellence.

FAILURE MODES FOR MANAGING FOR EXCELLENCE

Managing for excellence is critical for the institutionalization of Six Sigma to occur. This step specifically has to do with leadership. Four common failure modes include the following:

- Wrong people assigned to the projects.
- Leadership doesn't understand their role.
- No centralized training plan.
- Lack of understanding of Six Sigma resulting in poor support from Finance, IT, HR, Maintenance, and QC Lab.

These failure modes can be tracked directly back to a poor deployment. With great preliminary training of the company leadership, everyone will understand the importance of executing Six Sigma. The hallmark of any successful change initiative is getting the right people positioned in the right places. Integrating Six Sigma into the businesses is important to sustain the effective support of the support functions. None of these failure modes are hard to understand, but prevention must be considered in advance. But, the mark of superb leadership is making these difficult things happen.

Step 5: Sustain the Gains. Figure 16.5 shows a summary of Step 5, sustaining the gains. The two hardest things to do in any organization is (1) Start a new process and (2)

Stop the old process. That's why implementing a control system everywhere for every process is so important. There should be a control system for each Black Belt and Green Belt project. The process owner owns the control system. Part of the control system is to maintain a "project hopper," which is the repository for potential follow-on process improvement project. The control system undergoes regular reviews to verify the effectiveness of the Six Sigma system. Each project is reviewed at six months and one year after completion to ensure that the gains are still occurring. And, finally, the BB, GB, and Champion ranks are constantly refreshed.

Sustain the Gains

- Implement effective control plans
- Conduct regular training focused on the process
- Review the system effectiveness quarterly
- Continually identify and launch new projects

Figure 16.5 A summary of the actions associated with Step 5: Sustaining the Gains.

FAILURE MODES FOR SUSTAINING THE GAINS

Sustaining the gains is a critical step in both attaining and continuing the performance for which you are driving. The most difficult event in problem solving is not actually solving the problem. Rather, it is instituting the new change and making the change last. Six common failure modes include the following:

- No formal control plan system: Inadequate or incomplete project control plans.
- No process performance metrics in place.
- No process audit system in place.
- Control plans not implemented properly.
- Control plans not owned by process owner.
- System assumed to be working; no system reviews in place.

The Step 5 failure modes are oriented toward the lack of a definitive control system. Control systems are common, though I'd say not excellent, in manufacturing operations. In the transactional processes, there is usually little in the way of control systems or the systems are not complete. A combination of performance metrics and process audit systems are necessary to sustain the gains over the long term. Developing a clear method for implementing new control systems is the opportunity represented by Step 5.

IS YOUR DEPLOYMENT EFFECTIVE AND EFFICIENT?

The next Six Sigma assessment activity after assessing the leadership aspect of deploying Six Sigma is to assess the effectiveness or efficiency of your original deployment plan or the upcoming deployment plan. There are four stages to any Six Sigma deployment: (1) Plan and Assess; (2) Engage; (3) Execute and Practice; and (4) Internalize. Here is a summary of actions for each step:

- **Plan and Assess**
  - Create strategic alignment.
  - Define finance rule set to track results.
  - Executive interview process.
  - HR decisions.
  - Establish project pipeline.
- **Engage**
  - Executive Launch summit.
  - Communication plan in place.
  - First round of Champion sessions.
  - Resource plan for Six Sigma in place.
- **Execute and Practice**
  - Black Belt training launched.
  - Lean events activated.
  - Project mentoring initiated.
  - Green Belt training launched.
  - Project execution underway.
- **Internalize**
  - Master Black Belts developed.

- Strategy, HR, and rewards linked to Six Sigma.
- Train-the-trainer has begun.
- Institutionalize project pipeline process.

Is your culture ready to change? You may well ask the question as to whether your culture is ready for change. The following set of questions is useful in assessing the organization's readiness for change. Respond to these questions on a 1 to 10 scale (1=absolutely no and 10=absolutely yes).

Cultural Readiness to Change Scale

1. Is the organization structure relatively stable; that is, no dramatic changes soon?
2. Will the business leader make Six Sigma a top priority?
3. Does leadership have credibility and a history of successfully implementing company-wide improvement initiatives?
4. Will the business devote 10 percent of its resources to Six Sigma?
5. Will the business leadership team dedicate its top talent to Six Sigma?
6. Will members of the leadership team invest two days of their time to learn how to be credible Six Sigma Leaders?
7. Is the business leadership team open to actively sponsoring Six Sigma projects?
8. Is the organization comfortable working in teams?
9. Are decisions based upon analysis of relevant data at all levels in the organization?
10. Is work defined in terms of processes? Are key processes documented and accountabilities clear?

Based on the total score when summing from item 1 to item 10, a score of 75+ means your culture is probably ready for change. If the total score falls between 50 and 75, there may be some risk. If the total score is below 50, you would want to pay close attention to the culture either before deployment or within a few months after deployment.

IS EVERYTHING ALIGNED TO THE STRATEGY?

I've talked about alignment throughout this book. Aligning the external realities with the strategy, the financial targets, and the internal actions is the mantra. If your Six Sigma deployment is waning or you think there are risks to success before the deployment, check alignments.

Aligning Six Sigma Projects to Strategy. Chapter 9, "Committing to Project Selection, Prioritization, and Chartering," covered this topic extensively. Projects should be aligned to goals set around financial targets or operational metrics and then aligned to strategic thrusts. Using goal trees to depict the project, metric (Big Y), and strategic thrusts alignment is extremely visible, simple, and effective. Deploying the use of goal trees is a great method for ensuring proper alignment of projects. The projects should be aligned with the goals of each business in addition to the strategy. Figure 16.6 demonstrates the linkage from the strategic goals of improving working capital to the business goal to a project cluster to project number one. Success in project number one improves working capital for the business and working capital strategically.

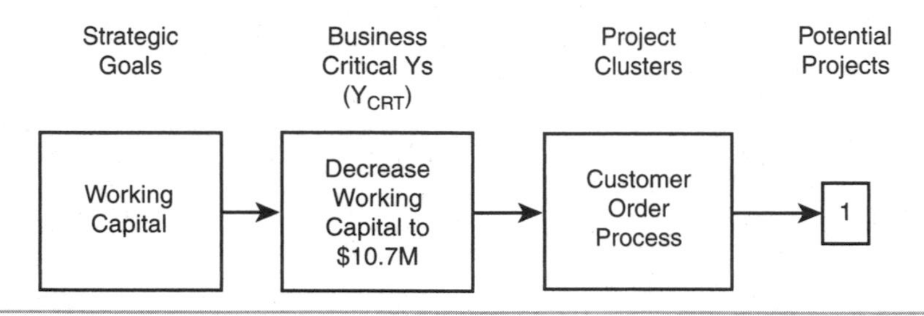

Figure 16.6 The linkage among strategic goals, business goals, and projects.

Are You Communicating Effectively? Effective communication within a Six Sigma deployment is essential. It's not a bad idea to assess your communication of Six Sigma and update your communication plan. The following factors are a great place to start:

- Who needs to be informed?
- What is the message?
- When does the communication need to occur?
- Why does the communication need to take place. . .the outcome?
- Who will inform them?
- What is the media?
- What is the rhythm?

Chapter 12, "Communicating the Six Sigma Program Expectations and Metrics," covers communication in detail. Review that chapter and then answer these questions, and you will assess (1) If you are ready to deploy or (2) Whether the current plan needs updating. Don't underestimate the effect of the communications plan on your change and your deployment.

Is the Finance Function Actively Participating? Because Six Sigma is focused on financial performance, the finance function within every business in a company leverages their knowledge, databases, and expertise to the entire Six Sigma process. CFOs are commonly major players in the deployment. The railroad company, CSX, had their CFO champion the entire process. The CFO at Rockwell International is heavily involved because of his positive introduction to Six Sigma while at AlliedSignal. The summary of the Finance roles follow:

- **Provide support throughout the Six Sigma projects**
  - Qualify project opportunities.
  - Identify leverage opportunities.
  - Validate process measurement proposals.
  - Verify projected dollar benefits as goals.
  - Characterize area of financial impact.
  - Provide ongoing project financial improvement evaluation.
- **Substantiate end-of-project savings/benefits**
  - After project completion.
  - Authentication of project financial results.

The financial function is to create and validate financial results. In Table 16.1, an estimate of the financial impact for a business's Six Sigma project plan is shown. The financial function estimates about a $120,000,000 impact on pretax income. That would be equivalent to starting a $1 billion business with no new capital.

Table 16.1 Forecast of the Financial Impact of Black Belt and Green Belt Projects for a Year

	Year 1 ($M)	Year 2 ($M)	Year 3 ($M)
Black Belts ($250K/Year)	$38	$38	$76
Green Belts ($75K/Year)	$22	$22	$44
Total	$60	$60	$120

Are You Tracking Your Program's Progress? Chapter 14, "Defining the Software Infrastructure: Tracking the Program and Projects," was spent talking about enterprise-wide tracking systems. How are you tracking the progress of your current or future program? Figures 16.7–16.9 show screenshots of a project-tracking system that tracks specific aspects of project work.

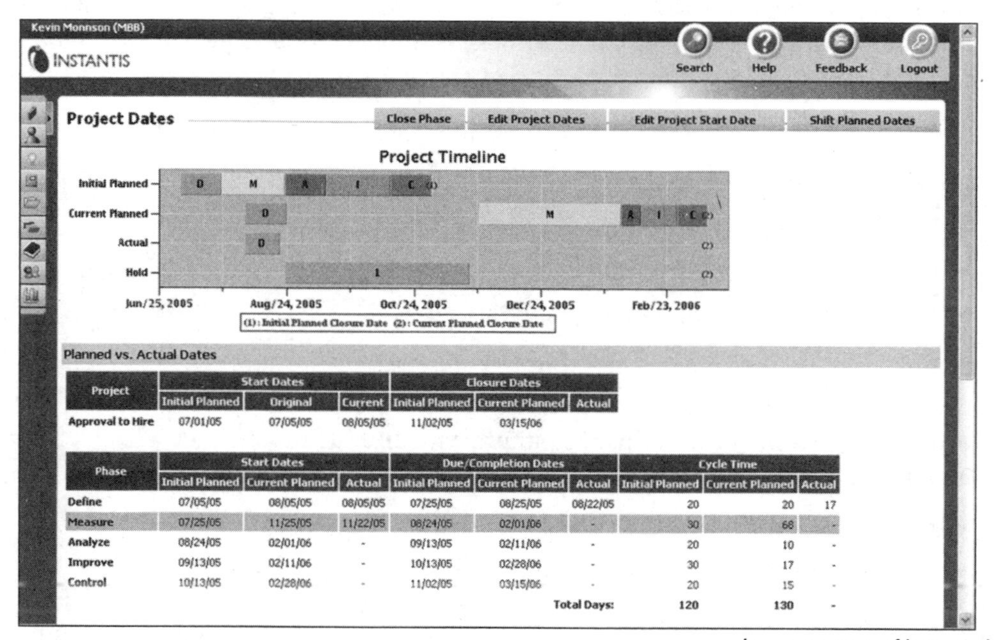

Image courtesy of Instantis, Inc.

Figure 16.7 Project Define-Analyze-Improve-Control workflow.

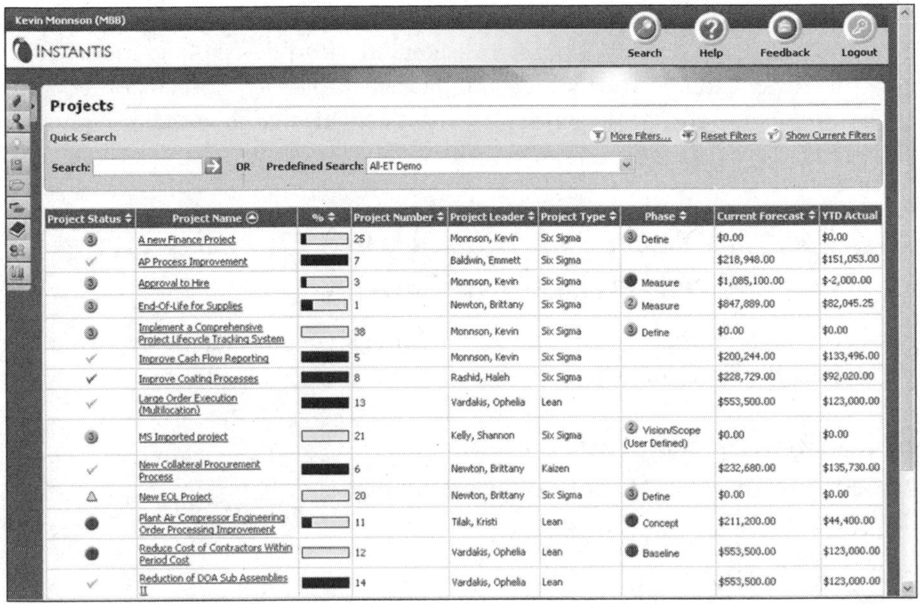

Project Status	Project Name	%	Project Number	Project Leader	Project Type	Phase	Current Forecast	YTD Actual
	A new Finance Project		25	Monnson, Kevin	Six Sigma	Define	$0.00	$0.00
	AP Process Improvement		7	Baldwin, Emmett	Six Sigma		$218,948.00	$151,053.00
	Approval to Hire		3	Monnson, Kevin	Six Sigma	Measure	$1,085,100.00	$-2,000.00
	End-Of-Life for Supplies		1	Newton, Brittany	Six Sigma	Measure	$847,889.00	$82,045.25
	Implement a Comprehensive Project Lifecycle Tracking System		38	Monnson, Kevin	Six Sigma	Define	$0.00	$0.00
	Improve Cash Flow Reporting		5	Monnson, Kevin	Six Sigma		$200,244.00	$133,496.00
	Improve Coating Processes		8	Rashid, Haleh	Six Sigma		$228,729.00	$92,020.00
	Large Order Execution (Multilocation)		13	Vardakis, Ophelia	Lean		$553,500.00	$123,000.00
	MS Imported project		21	Kelly, Shannon	Six Sigma	Vision/Scope (User Defined)	$0.00	$0.00
	New Collateral Procurement Process		6	Newton, Brittany	Kaizen		$232,680.00	$135,730.00
	New EOL Project		20	Newton, Brittany	Six Sigma	Define	$0.00	$0.00
	Plant Air Compressor Engineering Order Processing Improvement		11	Tilak, Kristi	Lean	Concept	$211,200.00	$44,400.00
	Reduce Cost of Contractors Within Period Cost		12	Vardakis, Ophelia	Lean	Baseline	$553,500.00	$123,000.00
	Reduction of DOA Sub Assemblies II		14	Vardakis, Ophelia	Lean		$553,500.00	$123,000.00

Image courtesy of Instantis, Inc.

Figure 16.8 Overall project progress.

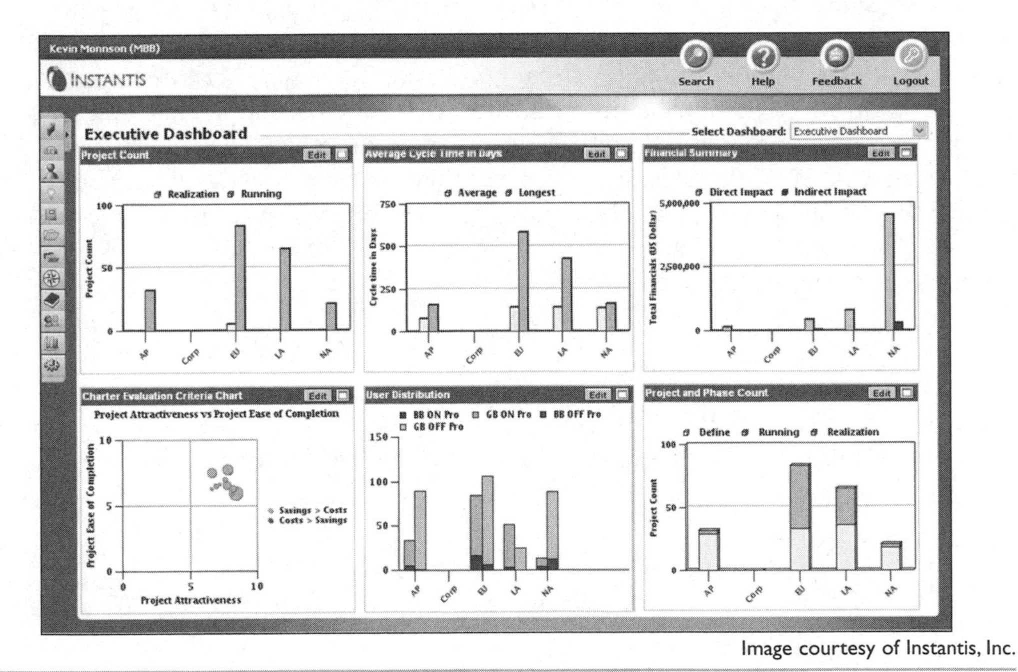

Image courtesy of Instantis, Inc.

Figure 16.9 Deployment progress.

There is a huge amount of benefit by tracking a change initiative across the enterprise from a centralized control point. Company leadership should be able to report the return on investment to the stockholder, Wall Street, employees, and customers. This process should be easy and comprehensive. The system also adds to the discipline the program needs to get institutionalized.

TIME TO EXTEND YOUR PROGRAM: USING DEPLOYMENT SUCCESSES TO GO FURTHER AND FASTER

Because Six Sigma is applicable across the enterprise, Six Sigma can be deployed into multiple arenas. There are quite a few ways to extend your Six Sigma program:

- Extend to other functional departments.
- Extend to involve more of your employee population.
- Extend your Six Sigma product lines.
- Extend to your supply chain.
- Extend to involve your customers.

Extend Six Sigma to Other Functional Departments. Any function within any business can benefit from Six Sigma. Anything of value the customer receives from us is an output of a process. Every function executes a set of processes. To be world class, every process must function effectively and be streamlined, efficient, and constantly improved. Any new product or service is the output of a development process. I facilitated a legal department that identified at least 45 processes they executed on a day-to-day basis. The list that follows shows only a small sample of projects for a variety of functions.

- **Finance Six Sigma Projects**
  - Financial Closing Activities, Cycle Time
  - Billing System Disconnects
  - Write-Offs—A/R, Bad Debts
  - Tax Credits—Sales and Use Tax, Recovery of Bad Debts, and so on
  - Cash Flow Management
  - Days Sales Outstanding
  - Price Increase
  - Accounts Receivable

    - Improve Payment/Collection Processes
    - Improve Audit Process Discovering Non-Compliance
    - Reduce Electronic Financial Transaction Costs
    - Reduce the Cycle Time of Reconciliation
    - Decrease Payment Processing Costs to Vendors/Suppliers
- **Human Resources and Administrative Six Sigma Projects**
    - Improved Processs for Transfer Assignments
    - Recruitment
    - Retention
    - Payroll Rejects/Administration
    - Sales Rep Turnover
    - Workers Comp Reduction
    - Union Effectiveness
    - Employee Training
    - Time and Attendance Efficiency
    - Medical/Dental Benefits
    - Leave Time Eligibility, Application, and Administration
    - Merit Pay Design
    - Ex-Pat Repatriation
    - Department Audit Process
    - Time Card Accuracy
    - Proprietary Information Destruction
    - Stock Option
    - Profit Sharing
    - Expense Report Cycle Time
    - Paperwork Reduction
    - Office Supply Acquisition
    - Office Machine Maintenance
    - Security Access and Control
    - Vacation Tracking and Control
    - Recycling

    - Absenteeism
    - Drug Testing
    - New Acquisition Integration
- **IT Six Sigma Projects**
    - Desktop Support
    - Disk Space Management
    - Functional Test Environment
    - Internet Response Time
    - Batch Job Scheduling Defect Reduction
    - Reduce Defect Leakage
    - Lab Hardware
    - ODC Entitlement
    - Reduce the Cost and Consumption Cycle Time
    - ERP—Migration, Optimization
    - Report Elimination
    - Computer Hardware/Software Acquisition
    - Orphan Data Reduction
    - Software License Process and Validity
    - Asset Management
    - Outsourcing Rationalization
    - Application Up Time
    - Software Development Life Cycle
    - Software Application Risk Management
    - Data Warehousing
- **Sales and Marketing, Communications, and Legal Six Sigma Projects**
    - Sales Force Effectiveness
    - Sales/Customer Interaction
    - Market Forecasting Process
    - Sales Projection
    - Proposal Creation
    - Case Work Effectiveness
    - Electronic Payables

- Debranded Marketers
- Margin Erosion in Sales Process
- Reduction of Dealer Price Variations
- Improve Timing of Sales Invoices
- Payment Transactions
- Corporate Goal Communication Process
- Public Perception Measurement
- Proprietary Information Control
- Communication-Devised Issuance and Control (Pagers, Beepers, Cells)
- Legal Documentation Efficiency
- Patent Control
- Product Liability
- Mergers Acquisition
- Consultant Contract Review
- Corporate Governance
- Call Center First-Time Resolution

Extend to Involve More of Your Employee Population. Early in most Six Sigma deployments, Six Sigma is applied to limited parts of the company. Many companies apply Six Sigma strictly to manufacturing operations. Therefore, only those folks involved in manufacturing have a chance to get involved with Six Sigma. Because only 3 to 5 percent of the population become Black Belts, that leaves a huge percentage with Six Sigma team membership as the only venue to Six Sigma.

The Green Belt role is a great way to widen participation in Six Sigma. A much larger percentage of the population can be Green Belts. Take a look at the first two years of the GE Six Sigma deployment. At the end of two years, GE had trained 4,000 Black Belts and 60,000 Green Belts, a huge difference in percentage of the total population trained.

When working at AlliedSignal, I had launched a Yellow Belt program, aimed initially at line operators on the manufacturing floor. This ended up being a very popular program, with hundreds of line operators receiving training in the same tools that the Black Belts use.

I also launched a training program for the first-line supervisors, another segment of the population brought to bear on Six Sigma. With a good needs assessment and some imagination, Six Sigma can be tailored to any number of subpopulations within your company. Developing unique applications of Six Sigma to unique populations is a great way to reinvigorate your program in the right way.

Extend to Include Additional Six Sigma Products and Services. The product portfolio for Six Sigma is huge. There will always be training applications that you won't be using. Taking a good look at your strategic gaps where Six Sigma is missing and plugging those gaps with new Six Sigma methods will maintain the momentum in your Six Sigma program. Chapter 11, "Selecting and Training the Right People," discusses the wide array of Six Sigma applications. By studying that portfolio and assessing where your impact is occurring, you can design the next phase of your Six Sigma deployment.

Adding Lean Enterprise to the Six Sigma initiative is one way to expand Six Sigma. Many companies start with manufacturing and move Six Sigma to product development and design. Some companies move Six Sigma through the companies in phases, with one emphasis at a time. Other, bolder companies drive Six Sigma in many different areas simultaneously.

Extend Six Sigma to Your Supply Chain. Many suppliers have noticed that when their customers launch Six Sigma, it's not long before the customers are requiring their suppliers to participate as well. Because suppliers are a major part of the value chain map, it only makes sense—manufacturing speed and quality are driven in large part by the efficiency and quality of their suppliers.

The idea here is that your supply chain management leaders aggressively start a Six Sigma program made available to the supply base. To receive this training, the suppliers must, in cooperation with the host company, commit to their own Six Sigma goals that tie to improving their performance for that particular customer (speed or quality). The hosting company provides the training and mentoring support from their own Master Black Belts and Black Belts. The host company will also provide leadership training for key suppliers. The point is, if your company develops a core competency in Six Sigma, why not provide that competency to key suppliers to optimize the entire value chain.

Align the Supplier Six Sigma program to incorporate corporate goals. . .assess whether the supplier's vision, customer requirements, critical processes, and improvement activities are parallel to yours. It is best to improve your internal process capability first before extending Six Sigma projects to suppliers. Make sure that all processes to which suppliers will be connected are optimized.

You must lead from the front if you expect your suppliers to take the Six Sigma interface seriously. When at Motorola, I ran a manufacturing line that produced electronic engine controllers with one customer, Ford. At that time (1988), Ford was pushing statistical process control (SPC). While we implemented a great SPC system, it was noted that, when touring Ford's factories, there was a conspicuous absence of SPC charts. That invalidated their threats around SPC. We simply did it because it was the right thing to do.

When connecting suppliers to your Six Sigma program, you must provide clear, actionable Voice of the Customer (VOC) to suppliers providing inputs to your organization's processes. If your organization does not know its own processes to a point where those requirements can be clearly communicated, suppliers will be navigating improvement projects without a compass. You will also lose credibility with your suppliers.

Extending Six Sigma to your supply base has a huge potential ROI. And it helps your suppliers become more competitive and willing to invest in improvement, making you look good. Launching a supplier program in Six Sigma will also help you focus on your own program to ensure that your program is the best it can be.

Extend Six Sigma to Your Key Customers. It may seem presumptuous to link your customers into Six Sigma, but it works if you do it right. Once mastered, Six Sigma becomes a great marketing tool. GE Health Care uses Six Sigma services as part of their CAT scan sales and other products. The AlliedSignal experience told us we could effectively market Six Sigma into the customer base. In one case, we moved from being completely de-sourced with one customer to becoming that customer's sole supplier.

The CEO of the company said it was simply because we could help him deploy Six Sigma. In other AlliedSignal businesses, we were able to get more volume and better prices by using Six Sigma as an additional service. We also got very close to those customers working with us. Our Black Belts worked in their factories and got direct voice of the customer.

As with suppliers, it's best to develop the Six Sigma core competency before striking out into the customer base. Using Six Sigma as an interface with your customers, you will engage them at a different level, and you will be able to share the benefits of Six Sigma.

- **Engage your customers**
  - Hear their voice
  - Create a relationship of trust
  - Partner to develop strategic growth plans
- **Shared Benefits**
  - Common language
  - Disciplined methodology for solving collective problems
  - Data-driven decisions
  - Embeds customer-thinking deep within both organizations

Six Sigma Success Factors. To successfully deploy Six Sigma, there are at least nine success factors to pay attention to and to assess when you are reinvigorating your program. The success factors are as follows:

1. Deployment plan—Chapter 3
2. Senior leadership training—Chapter 10
3. Strategic projects—Chapter 9
4. Development of technical experts—Chapter 11
5. Resources available and motivated—Chapter 13
6. Training and training plan—Chapter 11
7. Communications—Chapter 12
8. Project reviews—Chapter 9
9. Project tracking—Chapter 14

By reviewing the associated chapters in this book and identifying the gaps in your deployment, you can accelerate the rate of institutionalization of Six Sigma.

ASSESSING YOUR PROGRAM USING KOTTER'S EIGHT STAGES

As discussed in Chapter 3, "Six Sigma Launch Philosophy," the Harvard professor, John Kotter, has delivered a very workable and straightforward model for change. I will list his recommended steps with common failure modes seen in Six Sigma deployments, as follows:

1. **Establishing a Sense of Urgency**
   a. Failure modes:
      i. No sense of urgency created—no accountability.
      ii. Absence of senior leadership.
      iii. No commitment of leadership to understand change.
      iv. Complacency rules the day.
2. **Creating a Guiding Coalition**
   a. Failure modes:
      i. Champions have poor positional power.
      ii. No formal coalition established.
      iii. Poor commitment to up-front workshops and training.
      iv. No clear accountability for results.

        v. Guiding deployment with weak steering teams.
        vi. Not weeding out people against the change.

3. **Developing a Vision and Strategy**
    a. Failure modes:
        i. No vision developed—program is a training program.
        ii. Vision not linked to strategy.
        iii. No sense of urgency to support the vision.
        iv. Six Sigma not linked to strong results.
        v. Doing Six Sigma to check a box.
4. **Communicating the Change Vision**
    a. Failure modes:
        i. No communication plan—Six Sigma becomes a stealth program.
        ii. Communicated at upper levels but not at the lower levels.
        iii. Leadership not visible in their commitment and communication.
        iv. Under communicating.
5. **Empowering Employees for Broad-Based Action**
    a. Failure modes:
        i. No leadership commitment to intense workshops and training.
        ii. Little involvement in project selection.
        iii. No project tracking.
        iv. Six Sigma viewed as a nice training program.
        v. Little onsite project support given.
6. **Generating Short-Term Wins**
    a. Failure modes:
        i. No accountability established.
        ii. More than 12 months to achieve reasonable results.
        iii. No formal recognition ceremonies with clear presence of senior leadership.
        iv. No clear financial support to establish business impact.
        v. Launching too many projects at once.
        vi. Providing the first win too slowly.

7. **Consolidating Gains and Producing More Change**
   a. Failure modes:
      i. Program stagnates.
      ii. Internal experts not developed.
      iii. Program results not carefully tracked via metrics.
      iv. Six Sigma projects seen as extra work that detracts from day-to-day operations.
      v. Convincing the organization that they're done when they're not.
8. **Anchoring New Approaches in the Culture**
   a. Failure modes:
      i. Business as usual—if Six Sigma disappeared, no one would notice.
      ii. Promotions not linked to Six Sigma activities.
      iii. No development of MBB internal resources to support program.
      iv. Few changes in systems and leadership style from one wave of BBs to the next.
      v. Students consistently show up at training with poorly defined, small projects.

By reviewing Kotter's eight stages as an assessment, you will be able to quickly find holes in your Six Sigma deployment plan or your Six Sigma launch. Developing a core competency around Kotter's eight stages means you have set the stage for other change initiatives.

SIX SIGMA: THE INITIATIVE

Six Sigma has been one of the most successful change initiatives in decades mostly because it works. Understanding the deployment of change initiatives and driving one initiative to a successful conclusion is a valuable service to the future of any company. Larry Bossidy, in his book, *Confronting Reality*, gives four aspects of leading initiatives:

- Intense focus
- Hard work
- Tremendous time
- Endless physical and emotional energy

These are the behaviors that senior leadership exhibits in successful initiatives. But, nothing beats a great deployment plan based on preventing the common failure modes. The plan, after all, has to reach all levels of your company and motivate everyone to work as hard as you do to reach success. As a leader, you have to honestly answer these questions to plan your deployment.

- Do you know the initiative well enough to develop metrics to assess success?
- Are you willing to commit the resources to education?
- Do you have the focus and discipline?
- Do you have the stamina?
- Do you have the courage to confront those who stand in the way?

I went to school at Southwest Texas State University. SWT's most well-known alumnus was Lyndon B. Johnson, former President of the United States. After Lyndon retired from the Presidency, he honored our Student Senate at SWT by sitting in on one of our meetings. This was the meeting where a friendly coalition and I tried to pass a bill to eliminate the school football team and use the resulting financial savings for student sports. Considering the status that football has in Texas, needless to say I had to lay low for a while. Well, our bill lost, and that was that.

Lyndon made some summary remarks that went about like this. "I've been watching y'all try to do politics. You're not very good at it. Leading in the political theater is basic. First, you decide whether what you want to accomplish is right. If you decide it's right, then (the words of wisdom) you run over anyone that gets in your way!"

That's not the most humanistic approach for sure, but LBJ accomplished an amazing amount of change during his political career. That includes bringing electricity to rural Texas by teaming up with rival Republicans. His comment has always stuck with me. When I decided that what I wanted to do was the right thing to do, I heeded his advice and knocked over every barrier that blocked the way.

When launching Six Sigma, you first have to believe it's the right thing for the company's future. Once you decide Six Sigma is the right thing, then follow good form (Kotter's eight stages) and learn from those who came before you. A successful Six Sigma launch will have a positive impact on all your people and will be the one event in your career that you will remember fondly. Good luck, and I'll see you on the road to Six Sigma!

Appendix A
RFP Sample Format

Request for Proposal for Company X Six Sigma Deployment

TABLE OF CONTENTS

Item	Description
	Bidding Instructions
	Contractual Agreement
Attachment A	Statement of Work
Attachment B	Price Matrix

BIDDING INSTRUCTIONS

1.0 General Information.
1.1 Request for Proposal (RFP).
1.2 Cost Reimbursement.
1.3 Proposal Format and Number of Copies.
    1.3.1 Content.
1.4 **Proposal Due Date.** Your proposal is due ____________.
1.5 **Proposal Expiration Date.** Your proposal must be valid for a period of 60 days from the proposal due date specified in Paragraph 1.4 above. Company X reserves the right to negotiate changes, to reject any proposal or parts thereof, or to not make any award.
1.6 **Signature.** Your proposal shall be signed by an official authorized to bind your company to the order of magnitude contemplated.
1.7 News Releases.
1.8 Alternate Proposal.
1.9 **Contractor Survey.** To assess your ability to meet all proposed plans and commitments, Company X reserves the right to conduct surveys of your facilities, and the facilities of any subcontractors working on this project, at any time during the contractor selection process or during the period of contract performance.
1.10 **Oral Presentation.** The Contractor shall provide an oral presentation to the Company X team on _________, 2006 and _________, 2006. Details will be provided.
1.11 **Clarification.** All questions regarding this RFP shall be provided in writing via the Internet to __________@_____.com or faxed to ___-___-____ by ________, 2006.
1.12 **Amendments.** This RFP may be amended by Company X at any time. Contractors must acknowledge receipt of any amendment to this RFP by signing and returning the Amendment via fax within three (3) working days of amendment date.
1.13 **Contract Award.** It is anticipated that one award will be made by _______, 2006.
1.14 **Period of Performance.** Contract Award through ________, 2008.
1.15 **Contract Type.** Firm fixed rates.
1.16 **Evaluation.** Award will be made on a competitive basis to the Contractor submitting the “best value” proposal that is most advantageous to Company X. The evaluation shall occur in the following areas listed (in alphabetical order):

- Availability
- Cost
- Depth of Program

- Experience
- Flexibility

2.0 Business Operations and Technical Proposal.

2.1 **Executive Summary.** The Contractor shall summarize in this section why it should be awarded this business.

2.2 **Legal Entity.** The Contractor shall furnish the following information:

a. The legal name of its Company as it should appear on any order that may be issued.

b. The complete street address and mailing address (if different).

c. The legal form under which the Company does business (sole proprietorship partnership, corporation, etc.).

d. If the firm is a Division or Subsidiary, the name and address of its Parent Company.

e. The date upon and state in which the Company was formed or incorporated.

f. Whether the Company is classified as a
   - Large Business
   - Small Business
   - Woman-Owned Business
   - Minority Business Enterprise (MBE)
   - Foreign Business

2.3 Company Organization Chart.

2.4 Facilities/Capabilities.

2.5 Additional Capabilities and Excess Capacity.

2.6 **Experience.** The Contractor shall describe the number of years of experience it has in providing the services requested in this RFP.

2.7 **Company X Business.** What percent (%) of your sales is Company X business?

2.8 **References.** The Contractor shall provide a list by ________, 2006 of the names and phone numbers of

a. Three client references that Company X may contact for which the Contractor currently provides services requested in this RFP.

b. Three former clients that Company X may contact that have terminated or non-renewed the services included in this RFP. The Contractor shall also provide an explanation for the termination or non-renewal.

c. What Company X businesses in the United States have used your services? Please provide name, phone number, date, and description of the service.

d. List corporate clients with corporate deployment contracts that are active right now.

2.9 **Requirements.** The Contractor shall specifically include a statement that it will comply with all the requirements of Exhibit A Statement of Work. Exceptions, if any, must be stated in writing at the time your proposal in submitted, must be specific rather than general, and must be accompanied by a specific reason and alternative language.

2.10 Technical Requirements.

2.10.1 *Materials*

- Provide copies upon request of each of your Six Sigma materials for Champion, Master Black Belt, Black Belt, Green Belt, and Train-the-Trainer for manufacturing and transactional services for Company X's review. Describe how Company X can use this material.
- Describe how your team and materials will be tailored to Company X's requirements.
- Confirm that your material is available in both hard copy and electronic formats.
- How would you integrate current Company X quality and change initiatives into formal training material?
- Identify the foreign languages in which your materials are available.
- Describe the Company X Corporate global licensing arrangement for your materials.

2.10.2 *Training*

- Describe your training approach for Champion, Master Black Belt, Black Belt, Green Belt, and Train-the-Trainer for manufacturing and transactional environments.
- How do you ensure that the Trainer and the Consultant providing support are the same person to maintain continuity during the project?
- Explain how your coaching sessions are structured so Company X receives the best value for the service.
- Provide a listing of your Six Sigma course offerings.
- Describe your Train-the-Trainer program.
- Describe Six Sigma software and Internet tools.

2.10.3 *Experience*

- How many companywide Six Sigma deployments has your company managed? Describe them in general terms (manufacturing-chemical, manufacturing-discrete transactional). Summarize the results.

- How do you know your program is successful?
- Are you willing to defer payments until the value of your services are evidenced?
- What's your measurement system to ensure goals are met?
- What, and how, is your company held accountable for, and responsible for, during the period of performance of the contract?
- What is the typical time for a client to become self-sustaining?
- What are the typical financial returns on Company X's investment in your services?

2.10.4 *Consultants*

- What percentage of your trainers/consultants speak a foreign language? Please identify the language(s).
- Provide a summary, including educational background, type of industrial experience, and years of experience for manufacturing and transactional services consultants.
- Describe the process to be used with Company X for Company X consultant interviews.
- Confirm that you are willing to provide Trainers/Consultant support to facilitate common materials tailored to Company X's needs.
- Describe how you provide on-going project support.

2.11 **Confidential Information.**

2.12 Account Management.

- Describe your company's approach to customer service and how you would manage the Company X account.
- Identify the account manager and/or a project manager who will be assigned to the Company X account. This individual(s) must have the appropriate authority and responsibility to ensure all requirements of the contract are met.
- Describe how you will ensure that appropriate backup is available for the account manager, project manager, and trainers.
- What would your typical deployment to a Company X division consist of?

2.13 **Risks.** The Contractor shall identify the risks associated with this effort; describe the complexity of each risk; categorize them as Low, Medium, or High; and describe how it intends to mitigate the risks.

3.0 Price Proposal.

3.1 **Exhibit B, Price.** Contractor shall complete Exhibit B, Price Matrix and include the completed form in the proposal.

1. Licenses—Corporate Global. Identify what is included in this fee; e.g., software, materials, Facilitator Guide for internal trainers, etc. Please indicate the price of the software.
2. Training Materials for each category as shown.
3. Consultant fees for classes and on-going support.
4. Mark-up fees for independent contractors.

3.2 **Payment.** A milestone payment schedule will be developed and mutually agreed upon.

3.3 **Additional Terms and Conditions.** The following terms and conditions will be included in the contract issued for deployment of Six Sigma:

1. **Prohibitions.**
2. **Consultants.** Company X may review and approve all personnel providing services under this contract. This includes account manager and/or project manager and consultants. Contractor agrees not to assign work on Company X's premises to any of its employees who are not suitable to Company X. Contractor agrees to remove any individual from the project immediately if requested by Company X.
   The Contractor will provide Trainers/Consultant support to facilitate common materials tailored to Company X's requirements. These materials may not be Contractor materials.
3. **Account Management.** A designated Account Manager and/or Project Manager will be assigned to the Company X account and will have the appropriate authority and responsibility to ensure all requirements of the contract are met.
4. **Materials.** All materials shall be available in hard copy and electronic format and must be compatible with Company X software; e.g., MS Office, Windows, Lotus Notes, or Minitab. All materials shall also be available in foreign languages. The specific languages are TBD.
5. **License.** The Contractor shall grant to Company X and its present and future subsidiaries a permanent, irrevocable, royalty-free license to use electronic versions of the training materials and presentations (including any audio or video recording or other record of training presentations), together referred to as the "Training Materials," for Company X worldwide distribution for Company X internal use. Additionally, Contractor understands and acknowledges that Company X may disclose the Training Materials, not for resale, to

third parties (such as suppliers, customers, and distributors) with whom Company X has a commercial relationship; Contractor agrees that such distribution does not violate Company X's license hereunder and is included in the license fee paid by Company X.

6. **Statistical Software.** Company X will work with the Contractor to determine the primary statistical package for the corporation.
7. **Key Personnel.** The Contractor shall notify the Company X Program Manager in writing 30 days prior to making any change in Key Personnel. Key Personnel are defined as follows:
   a. Individuals who are designated as Key Personnel by written agreement of Company X and Contractor.
   b. The Account Manager, ___________, is hereby designated as Key Personnel.

   Notwithstanding any forgoing provisions, the Key Personnel shall be assigned to perform services for the period specified in this agreement unless the Contractor has demonstrated to the written satisfaction of Company X's project leaders that the qualifications of the proposed substitute personnel are equal to or better than the qualifications of the person being replaced. The provisions of this paragraph shall not apply if the Key Personnel leave Contractor's employ, or are no longer available due to circumstances beyond Contractor's control.
8. **Hiring Practices.** During the term of this agreement (and for a period of three years thereafter), Contractor agrees not to solicit or hire any Company X employees trained in Six Sigma. Contractor shall agree to instruct its employees, and its subcontractors, not to actively solicit Company X for employment opportunities during the term of this Agreement.
9. **Payment Schedule.** Company X and the Contractor will mutually agree to a milestone payment schedule, which will be incorporated and made a part of this agreement by reference.
10. **Price.** All consultants' fees and materials costs for Company X worldwide shall be as shown in Exhibit B, Prices. No other prices will be allowed.
11. **Global Agreement.** This RFP applies globally and will include Six Sigma deployment for the United States.
12. **Technical Contract Worker Agreement.** Each individual consultant providing Six Sigma training to Company X will be required to sign a Technical Contract Worker Agreement prior to starting work at Company X.

Exhibit A
Statement of Work

OBJECTIVE AND GOAL

To deploy the Six Sigma training process as a global corporate initiative for its business units and staff groups.

Company X's goal is to provide a standard Six Sigma methodology with common materials to ensure uniformity throughout the corporation.

REQUIREMENTS

All Six Sigma training will use ______ statistical software. The Contractor shall provide Six Sigma Training for the following deployment:

1. Champions—approximately ____ Start ____, 2006.
2. Master Black Belts—approximately ____ Start ____, 2006.
3. Black Belts—___-___ Start ____, 2006.
4. Green Belts—___-___ Start ____, 2006.

ACCOUNT MANAGEMENT

The Contractor shall provide a dedicated account manager and/or project manager, subject to Company X approval, to coordinate and manage the Company X account. This individual(s) must have the appropriate authority and responsibility to ensure all requirements of the contract are met.

DELIVERABLES

Customized training materials—hard copy and electronic for Champions, Master Black Belts, Black Belts, and Green Belts.

Facilitators Guide for the internal trainers—hard copy and electronic.

Monthly project reports.

Exhibit B
Price Matrix

Description	Price
Corporate Global License Fee	$
Executive Workshops (2 × 2-Days)	$
Division Workshops (4 × 2-Days)	$
Champions Workshops (8 × 3-Days)	$
Black Belt Classes (20 × 20-Days)	$
Green Belt Classes (30 × 10-Days)	$
Consulting fees—on-going support	$
Green Belt Training Material Package	
(4 × 4-Days)	$
Master Black Belt Classes (2 × 25-Days)	$

OTHER SERVICES

Please list below any other services required that are not shown above. Include the description and the price.

B About the Contributors

Michael Brennan is an executive director for Sigma Breakthrough Technologies, Inc. (SBTI). Brennan specializes in strategic planning, executive training in Six Sigma, new product planning, product development, and Six Sigma deployment. He previously was vice president of product development at Black & Decker Power Tools and the Life Fitness Division of Brunswick. He is an accomplished leader and change agent with a track record of delivering world-class results in global markets with strategic planning, innovative products, concurrent product development, manufacturing startup, and effective product marketing. Mike earned his MBA and M.S. in Electrical Engineering from the University of Louisville, Kentucky.

Joyce A. Friel is president of Peak Performance Consulting Corporation. With more than 25 years' experience in organizational development, human resources, and process improvement consulting, Friel serves as a business leader, strategist, facilitator, catalyst, and developer. Her applications experience comes from leading change in both Fortune 500 and smaller organizations, including Eastman Kodak, IBM, and Kodak Polychrome Graphics. As a visionary leadership strategist and catalyst for change, Friel concentrates on strategic planning, measurement and goal alignment; change management; leadership development, group process design, and facilitation. Joyce is an active member of Arizona Business Leaders, Scottsdale Chamber of Commerce, Society for Human Resource Management (SHRM), Arizona SHRM Consultant's Forum, International Association of Facilitators, National Organizational Development Network, and Executive Women's Golf Association. Friel has served on numerous community and educational advisory boards and is an adjunct faculty member of the Graduate School of the College of Business and Professional Studies at Grand Canyon University.

Joe Ficalora is currently the vice president of technology at SBTI. In this role, he is responsible for the methodologies, content, and technical approaches SBTI uses. Joe developed and designed the highly acclaimed SBTI Master Black Belt Program. He has worked, and is certified as, a Black Belt and Master Black Belt. He has consulted with clients in the medical devices, beverage, health care and food packaging, electronics, metal, glass, and plastics manufacturing industries. He has mentored, designed, and taught workshops to Executives, Champions, Master Black Belts, Black Belts, and Green Belts in DFSS, Six-Sigma in Manufacturing, and Transactional Business projects worldwide. Ficalora also holds several patents in lasers and optical devices, and is an active member in IEEE and ASQ.

Dino Hernandez is director of sales and marketing for SBTI. He leads the marketing and sales teams and champions SBTI's Marketing and Sales for Six Sigma products. With over 15 years' experience in management and business leadership, Hernandez has led businesses in Europe, Asia, and the U.S. In addition to his broad experiences in marketing, sales management, and product development, he is a certified Six Sigma Black Belt, with expertise in both the DMAIC and DFSS roadmaps. He has an MBA from the University of Texas and a BSEE from the University of Oklahoma.

Roger Hinckley is director of international operations for SBTI. He enhances SBTI's international activities by developing international joint ventures, partnerships, and affiliations; working with International JV leadership deployment planning and training; developing an operational infrastructure; and reporting with managing translations of SBTI Intellectual Property (IP) materials to facilitate training and testing in English, French, German, Spanish, Dutch, Italian, Japanese, Chinese, and Korean. Hinckley has also contributed to market analysis of market segments for Six Sigma applications, client pitch and negotiation with relevant financial modeling, executive material customization, deployment modeling, and coaching support.

Daniel M. Kutz is president of SBTI. He oversees business development, marketing, sales, and day-to-day operations. He previously served as vice president of operations and was responsible for all internal operations, expanding sales and marketing efforts, overseeing project managers, revising the leadership training curriculum, and expanding the regional training program. Earlier, as an SBTI master consultant, Kutz taught and consulted with client companies worldwide in Six Sigma methods and productivity improvement. He was the project manager coordinating the Six Sigma deployment efforts in Eaton Corporation, Cummins, Cooper-Cameron, Entergy, and BASF Automotive Coatings. He holds a B.S. in Electrical Engineering from Marquette

University and an M.S. in Manufacturing Systems and an M.B.A. from the University of Texas at Austin.

Kristine Nissen is a senior consultant with SBTI and is currently deploying process improvement methods to clients around the world. Ms. Nissen has successfully implemented Lean and Six Sigma techniques within global logistics and distribution, chemical, manufacturing, medical device, and service companies ranging from $400 million to $40 billion in annual revenue. She codeveloped SBTI's Transactional Green Belt and Black Belt curriculums and the Six Sigma Process Design methodology that enables companies to rapidly develop and deploy new processes and services that meet customer needs. Ms. Nissen has experience in all areas of business process management and improvement, with demonstrated expertise in Six Sigma, Lean, project management, Total Quality Management (TQM), DFSS, product cost reduction, and statistical modeling. She is a certified Master Black Belt.

Randy Perry is a senior-level executive for SBTI. He has extensive experience in P&L, global product management, marketing, Six Sigma, strategic planning, personnel coordination, and budgeting for global process operations. Perry is a Master Black Belt in Design, Operations, and Transactional Six Sigma methods, with experience in project implementation for various multinational corporations. Perry has worked with clients such as Seagate, Tyco, Celanese, Eastman Chemical, BASF, Sequa Chemical, WR Grace, TRW, Lincoln Electric, FMC, Johnson Wax, and Osram Sylvania.

Richard R. Scott is master consultant and executive director for SBTI. He manages the SBTI Master Black Belt Program, which trains and develops superior Six Sigma leaders; highly knowledgeable mentors and technical resources for Champions, Black Belts, and Green Belts; and excellent instructors, communicators, and future business leaders. Scott is an expert in Six Sigma for Operations, DFSS, Marketing for Six Sigma, project management, TQM, and statistical modeling.

Debby Sollenberger is vice president of business development for SBTI. She brings her experience as a General Electric executive to consult with senior leaders of SBTI's current and future clients. Sollenberger is a General Electric-certified Black Belt and Master Black Belt. She has designed, developed and delivered training in DMAIC, Design/Measure/Analyze/Design/Verify (DMADV), and Black Belt Leadership Training.

Dr. Ian Wedgwood is an executive director for SBTI, responsible for working extensively with clients, consultants, and home-office staff to ensure client success and continued business development. Dr.Wedgwood has led a number of deployments in industries as diverse as electronics, engineered materials, medical devices, chemicals, and healthcare, and has trained and mentored numerous Executives, Champions, and Belts in DFSS, Six

Sigma, and Lean. Dr.Wedgwood has a strong product development background and codeveloped SBTI's Lean Design, Lean Sigma, and Healthcare methodologies and curricula. He holds a Ph.D., and a first-class honors degree, in applied mathematics from Scotland's St. Andrew's University.

Sigma Breakthrough Technologies, Inc. (SBTI), a corporation with 28 employees and over 50 expert consultants, is based in San Marcos, Texas (only 25 minutes south of Austin, Texas). Specializing in Lean and Six Sigma deployments, the CEO and executive team has over 100 years of combined process-improvement and product-development expertise. SBTI has been delivering superior results through Six Sigma and Lean for customers since May 1997. SBTI executive directors and master consultants have a minimum of 10 years' industry experience, but many have 25 years' experience or more. Executive directors for SBTI have actually led program deployments inside a corporation and/or as an outside project manager. SBTI has worked with over 50 corporate clients. Find out more about SBTI at http://www.sbti-hq.com/.

Peak Performance Consulting Corporation is an organizational development company. With intentionally planned efforts specifically targeted to client needs, Peak Performance increases the client organization's effectiveness by increasing leadership capability, streamlining and defining effective processes, facilitating the development and use of balanced strategic plans, and regularly monitoring metrics. Collectively, these and other intentionally planned activities yield a shift in the organization's culture and results. Peak Performance works with companies to assess and design business solutions that deliver results, with customers ranging from small businesses to large Fortune 500 companies located throughout the U.S. and Canada. Find out more about Peak Performance Consulting Corporation at www.Peakperfromancecorp.com.

Instantis, Inc., provides on-demand software to manage CXO-mandated initiatives that improve enterprise financial performance. Leading global corporations, such as Credit Suisse, McKesson, Xerox, Verizon, and dozens of others, use Instantis software to manage initiatives such as Six Sigma, Lean, Operational Excellence, New Product Development, and more. Instantis's flagship EnterpriseTrack product enables CXOs to assure that teams convert their strategic priorities into visible execution in the form of structured, trackable, manageable projects. With unmatched configurability, ease of deployment, ease of use, ease of administration, and a comprehensive fourth-generation feature set, EnterpriseTrack is in production deployment at more Global 2000 companies than any other comparable product. Instantis customers span multiple industries—such as manufacturing, healthcare, services—and include market leaders. Find out more about Instantis at www.instantis.com.

Index

A

ABB (Asea, Brown, Baveri), 26
accountability
of projects, 179, 311
of Six Sigma providers, 58
The Alchemy of Growth (Baghai), 72
Alder, Jim, 46, 126
alignment
reinvigorating Six Sigma deployment, 341-346
Six Sigma for, 6-7
AlliedSignal, 26, 37-38
anchoring change initiatives (deployment architecture stage), 52-53, 304-305
Asea, Brown, Baveri (ABB), 26
assessments. *See also* change initiatives, reinvigorating
example, 326-328
with Kotter's philosophical deployment architecture, 352-354
of Six Sigma deployment, 331-333

B

Baghai, Mehrdad, 72
BBs. *See* Black Belts
Belts. *See* Black Belts; Green Belts; Master Black Belts
Bennis, Warren, 138
best practices, sharing, 296
big picture clarification. *See* linking internal activities
Black Belts (BBs), 8, 258
certification requirements, 144-146, 272-274
Chemical Design for Six Sigma training programs, 216-218
defined, 11
financial expectations, 36
full time versus part time, 260-262, 316
hiring, 259-260
K-Sigma Black Belt training program, 220-225
number to train, 228-230, 262-263
Operational Black Belt training program, 218-220

as part of infrastructure, 133-135
percentage to be trained, 36
project drivers, 249
responsibilities of, 10-11
retention and career planning, 279-281
succession planning, 271-272
talent selection, 259-268
time dedicated to projects, 230
training programs, 225
Bossidy, Larry, 3, 6, 17, 22, 26, 41, 43, 45-46, 49, 56-57, 69, 101, 107-108, 118-121, 236, 285, 330, 354
bottom-line expectations, 91-99
Bottom-Up prioritization matrix, 170-171
bottom-up projects, 10, 161-165
Breen, Ed, 6, 108
broad-based change empowerment (deployment architecture stage), 50
Burhnam, Dan, 6
business model, components of, 17-23, 69-70
business process reengineering. *See* reengineering
Business Team Workshop, 190-195
customizing, 199-204
business unit deployment, division deployment versus, 109-110
business units, Deployment Champions for, 126-128

C

capacity productivity (C-P), 87
career planning for Belts, 279-281
CEO, as part of infrastructure, 120-123
certification requirements, 144-146
for Belts, 272-274
Champion Workshops, 195-199
customizing, 199-204
Champions
Deployment Champions, 46
Initiative Champions, 45-46
responsibilities of, 10-11
change initiatives
anchoring (deployment architecture stage), 52-53, 304-305
connecting Six Sigma with, 106-108
driving change, importance of, 3
reinvigorating, 329-331
deployment efficiency, 340-341
extending Six Sigma, 346-352
leadership roadmap, 334-340
program assessment, 331-333
quality dimensions, 333-334
strategic alignment, 341-346
Six Sigma as, 6
steps to success, 354-355
change readiness, 341
Charan, Ram, 3, 17, 69
chartering projects, 176-182, 310
charters, defined, 27
Chemical Design for Six Sigma training programs, 216-218
clarifying big picture. *See* linking
Collins, Jim, 7, 42
communication (deployment architecture stage), 48-49, 240-241
communication plans, 235-236
creating custom, 238-240
elevator speeches, 250-252
example of, 246-249
Human Resources (HR) role in, 281-284
importance of, 322, 342
matrix for, 236-238
media for, 244-245
message presentation, 241-244
topics for, 252
communications department, extending Six Sigma to, 348
complacency, sources of, 42
complexity of message in communications matrix, 236-238
Confronting Reality (Bossidy and Charan), 3, 17, 46, 69, 119-120, 330, 354

consolidating gains (deployment architecture stage), 51-52, 303-304
consultant experience, differentiating Six Sigma providers, 65-66
consultants. *See* providers
contracts with Six Sigma providers, 67-68
control plans for sustaining performance, 324
COPQ (cost of poor quality), 87, 163
corporate deployment history, differentiating Six Sigma providers, 64
corporate history, differentiating Six Sigma providers, 64
cost of poor quality (COPQ), 87, 163
costs. *See* ROI (return on investment)
Covey, Stephen, 6, 56
crisis identification, 42
Critical Ys. *See* goals; operational metrics; outputs
Crystal Ball (simulation software), 299
culture of organization
  readiness for change, 341
  Six Sigma and, 37
customers, extending Six Sigma to, 351
customization
  differentiating Six Sigma providers, 64
  of workshops, 199-204

D

decision matrix, selecting providers, 61-63
defects per unit, 93-97
deployment
  communication plans. *See* communication plans
  components of, 38-40
  efficiency assessment, 340-341
  extending Six Sigma, 346-352
  Human Resources (HR) role in, 255-259, 281-284
    Belt retention and career planning, 279-281
    position profile example, 268-274
    recognition and rewards program, 274-279
    talent selection, 259-268
  Kotter's philosophical deployment architecture, 41, 352-354
    anchoring change initiatives, 52-53, 304-305
    communicating change vision, 48-49, 240-241
    consolidating gains, 51-52, 303-304
    employee empowerment, 50
    guiding coalition, 45-47
    sense of urgency, 42-45
    short-term gains, 51
    stages of, 41
    vision and strategy, 47-48
  scope of, 105-106
    business unit versus division deployment, 109-110
    connecting Six Sigma with other initiatives, 106-108
    domestic versus global deployment, 110-111
    integration of Six Sigma programs, 112-113
    integration with Lean Enterprise, 111-112
    pilot versus full deployment, 108-109
    scheduling events, 113-114
  success factors, 351
  timing, 40-41
Deployment Champion Workshop, 195-199
  customizing, 199-204
Deployment Champions, 46
  as part of infrastructure, 126-128
deployment letter example, 282-284
Design for Six Sigma (DFSS), 39, 113
  training programs, 215-218
desktop management software solutions, 298-300
DFSS. *See* Design for Six Sigma
differentiating providers, 63-66

discipline, Six Sigma for, 7-8
division deployment, business unit deployment versus, 109-110
DMAIC roadmap, steps in, 35
domestic deployment, global deployment versus, 110-111
DPU (defects per unit), 93-97
driving change, importance of, 3

E

efficiency. *See* bottom-line expectations
The 8th Habit (Covey), 6
"elevator speeches," 47, 250-252
emotional content of message in communications matrix, 236-238
employee empowerment (deployment architecture stage), 50
employees
  extending Six Sigma participation, 349
  versus independent contractors, differentiating Six Sigma providers, 65
enterprise management software solutions, 287-290
  packages for, 297-298
  requirements, 290-297
entitlement, 30-32
  defined, 161
  establishing, 310
evaluations of training plans, 230-234
Excel (spreadsheet software), 299
executing software solutions, 300-301
executive interviews, 199-204
Executive Team, as part of infrastructure, 120-123
Executive Team Workshop, 186-190
  customizing, 199-204
expanding software solutions, 301-302
expectations. *See also* financial targets
  bottom-line expectations, 91-99
  importance of defining, 90
  participation expectations, 100-101
  top-line expectations, 99-100
external realities
  assessing, 18-19
  crisis identification, 42
  linking internal activities with, 151, 159-160, 309-310
  reinvigorating change initiatives, 330
  strategic planning and, 69-85

F

failure modes
  in Kotter's philosophical deployment architecture, 352-354
  management support, 338
  process improvement plans, 337
  project selection, 335
  sustaining performance, 339-340
  training plans, 336
filters. *See* project filters
financial function
  extending Six Sigma to, 346
  importance of, 343
financial impact, tracking, 322
financial metrics. *See* financial targets
financial support, as part of infrastructure, 139-141
financial targets. *See also* expectations; money
  bottom-line expectations and, 91-92
  metrics and, 87-90
  reinvigorating change initiatives, 330
  setting, 19, 88-90
  Six Sigma processes and, 19-22, 73-75
  strategic planning and, 69-85
financial tracking programs, 294
Financial Workshop, 206
Fisher, George, 101
flexibility of software solutions, 296-297
Friel, Joyce, 236
full deployment, pilot deployment versus, 108-109
full time Belts, part time versus, 260-262, 316

G

gain-sharing agreements, 67
Galvin, Bob, 122-123, 245
gap analysis, 79
GBs. *See* Green Belts
GE (General Electric), 26
global deployment, domestic deployment versus, 110-111
global presence, differentiating Six Sigma providers, 66
goals, entitlement and, 30-32. *See also* financial targets
Good to Great (Collins), 7, 42
Great Groups, 138
Green Belts (GBs), 8, 259
  certification requirements, 144-146, 272-274
  defined, 11
  financial expectations, 36
  full time versus part time, 260-262, 316
  hiring, 259-260
  number to train, 228-230, 262-263
  as part of infrastructure, 135-137
  percentage to be trained, 36
  position profile example, 268-270
  responsibilities of, 10-11
  retention and career planning, 279-281
  talent selection, 259-268
  time dedicated to projects, 230
  training programs, 225
guiding coalition (deployment architecture stage), 45-47

H

Hammer, Michael, 57
handbook example, 325-326
Harry, Mikel, 267
head count reductions, 92
Hill, Bill, 11
hiring Belts, 259-260
history of Six Sigma, 25-26
Horizon 1 (strategic planning), 72-73
Horizon 2 (strategic planning), 72
Horizon 3 (strategic planning), 72
Human Resources (HR)
  extending Six Sigma to, 347
  role in Six Sigma deployment, 255-259, 281-284
    Belt retention and career planning, 279-281
    position profile example, 268-274
    recognition and rewards program, 274-279
    talent selection, 259-268
  support, as part of infrastructure, 141-142
Human Resources Workshop, 207

I

i-solutions (enterprise management software), 298
iGrafx (desktop management software), 300
independent contractors versus employees, differentiating Six Sigma providers, 65
industry experience, differentiating Six Sigma providers, 64
infrastructure, 117-119, 146-147, 257
  Black Belts (BBs), 133-135, 258
  CEO and Executive Team, 120-123
  certification requirements, 144-146
  Deployment Champions, 126-128
  financial support, 139-141
  Green Belts (GBs), 135-137, 259
  Human Resources support, 141-142
  Initiative Champions, 123-126
  Master Black Belts (MBBs), 130-133, 258
  Project Champions, 128, 130
  project team members, 138-139
  project tracking system, 142
  steering teams, 142-143
Initiative Champions
  defined, 11
  as part of infrastructure, 123-126
  selecting, 45-46
initiatives. *See* change initiatives

inputs
financial metrics and, 19-22
strategic planning and, 73-75
Instantis (enterprise management software), 297
integration
Six Sigma and Lean Enterprise, 111-112
Six Sigma programs, 112-113
intellectual property (IP), 58
differentiating Six Sigma providers, 65
internal activities
linking, 22-23
with external realities, 151, 159-160, 309-310
reinvigorating change initiatives, 330
internalization of training, 226-227
interviews, executive interviews, 199-204
IT department, extending Six Sigma to, 348

J–K

JMP (statistical analysis software), 299
Johnson, Lyndon B., 355

K-Sigma Black Belt training program, 220-225
Kearney, A.T., 56
knowledge management, 296
Kotter's philosophical deployment architecture, 41, 352-354
anchoring change initiatives, 52-53, 304-305
communication change vision, 48-49, 240-241
consolidating gains, 51-52, 303-304
employee empowerment, 50
guiding coalition, 45-47
sense of urgency, 42-45
short-term gains, 51
stages of, 41
vision and strategy, 47-48
Kotter, John, 6, 41
Kouzes, James, 8
Krugman, Herb, 247

L

labor savings, reducing head count, 92
Labovitz, George, 70, 309
leadership, principles of, 120
The Leadership Challenge (Kouzes and Posner), 8, 120
leadership development, Six Sigma as, 8-9
leadership roadmap, 303
management support, 319, 321-323, 337-338
process improvement roadmaps, 318-319, 336-337
project selection, 151-157, 307-311, 334-335
steps in, 150, 306-307
sustaining performance, 323-325, 339-340
training plans, 312-317, 335-336
leadership workshops. *See* workshops
Leading Change (Kotter), 6, 41
Lean Enterprise
integration with Six Sigma, 213-215
K-Sigma Black Belt training program, 220-225
Six Sigma integration with, 111-112
training portfolios, 211-213
Lean Sigma, 39
training portfolios, 211-213
legal department, extending Six Sigma to, 348
linking internal activities, 22-23
with external realities, 151, 159-160, 309-310
Linsenmann, Don, 124
Little, Arthur D., 56
Lucas, Wes, 6

M

MAIC (Measure, Analyze, Improve, Control) roadmap, 113
management support
failure modes, 338
importance of, 319, 321-323, 337-338
Mankins, Michael, 69

manufacturing processes
K-Sigma Black Belt training program, 220-225
Operational Black Belt training program, 218-220
marketing department, extending Six Sigma to, 348
Master Black Belts (MBBs), 8, 258
certification requirements, 144, 272-274
defined, 11
financial expectations, 36
full time versus part time, 260-262, 316
hiring, 259-260
number to train, 228-230, 262-263
as part of infrastructure, 130-133
percentage to be trained, 36
responsibilities of, 10-11
retention and career planning, 279-281
talent selection, 259-268
time dedicated to projects, 230
training programs, 225-227
materials review process, 204
MBBs. *See* Master Black Belts
McDonald, Frank, 46, 126
McNerney, Jim, 6, 9, 18, 43, 46, 108-109, 122, 126, 206, 285
measurable goals. *See* financial targets
media for communication plans, 244-245
message complexity, in communications matrix, 236-238
message emotional content, in communications matrix, 236-238
message presentation, in communication plans, 241-244
methodology of Six Sigma, 10-12
metrics
bottom-line expectations and, 91-92
defects per unit, 93-97
financial targets and, 87-90
participation expectations and, 100-101
rework, 99
scrap, 97-98
software for tracking. *See* software (for tracking metrics)
top-line expectations and, 99-100
Microsoft Office suite (desktop management software), 299
milestones
list of, 105-106
project selection, 158
scheduling Six Sigma deployment events, 113-114
Minitab (statistical analysis software), 299
money
delivering, 24-25, 35-36
finding, 24, 28-34
focusing on, 24-28
Motorola, 25, 149-150

N–O

Nordelli, Bob, 6
Norris, Paul, 6, 48, 101, 159, 309

operating activities, 22
Operational Black Belt training program, 218-220
operational metrics, prioritizing project clusters, 152-157
Operational Six Sigma, 39, 112-113
operations roadmap. *See* DMAIC roadmap
opportunity for success, identifying, 42
organizational culture
readiness for change, 341
Six Sigma and, 37
organizational priorities, polls about, 7
Organizing Genius (Bennis), 138
outputs
financial metrics and, 19-22
strategic planning and, 73-75

P

part-time Belts, full-time versus, 260-262, 316
participation expectations, 100-101
partnership with Six Sigma providers, 59-60

PDBBs (Product Development Black Belts), talent selection, 266-268
performance
 expectations, 297
 sustaining, 323-325, 339
  failure modes, 339-340
pilot deployment, full deployment versus, 108-109
planning software solutions, 300
polls, organizational priorities, 7
Poses, Fred, 6, 31, 42-47, 49, 58, 101, 120-122
position profile example, 268-274
Posner, Barry, 8
The Power of Alignment (Labovitz and Rosansky), 70, 149, 309
PowerPoint (presentation software), 299
PowerSteering Software (enterprise management software), 298
prioritizing
 project clusters with operational metrics, 152-157
 projects, 149-150, 165-175, 310
  Bottom-Up matrix, 170-171
  Motorola example, 149-150
  project filters, 172-175
  Top-Down matrix, 166-170
problem-solving ability with Six Sigma, 12
process baseline
 defined, 161
 establishing, 310
process entitlement
 defined, 161
 establishing, 310
process improvement in Six Sigma, 19-23, 73-75
process improvement roadmaps, 318-319, 336-337
processes
 entitlement, 30-32
 importance of, 309
 money generation. *See* money
 Six Sigma forms, 39
 value mapping, 28-30
procrastination costs, 44
Product Development Black Belts (PDBBs), talent selection, 266-268
productivity. *See* bottom-line expectations
profit margins, analyzing, 32-34
program assessments. *See also* change initiatives, reinvigorating
 with Kotter's philosophical deployment architecture, 352-354
 of Six Sigma deployment, 331-333
program reviews, 59
program tracking. *See* software (for tracking metrics)
Project (process management software), 299
Project Champion Workshop, 195-199
 customizing, 199-204
Project Champions
 defined, 11
 as part of infrastructure, 128, 130
project clusters, prioritizing with operational metrics, 152-157
project drivers for Black Belts (BBs), 249
project filters, 172-175
project prioritization and selection roadmap, 157-160
project results. *See* results
project team members, as part of infrastructure, 138-139
project tracking system, as part of infrastructure, 142
projects
 accountability for, 179, 311
 advantages of, 182
 Bottom-Up projects, 161, 163-165
 chartering, 176-182, 310
 continuous identification of, 324
 financial impact of, 24

idea development and selection, 292
managing, 292-295
performance expectations, 297
prioritizing, 149-150, 165-175, 310
Bottom-Up matrix, 170-171
Motorola example, 149-150
project filters, 172-175
Top-Down matrix, 166-170
for profit margin improvements, 32-34
reasons for lack of success, 182-183
results tracking, 295-296
reviewing, 179-182, 321-322
selecting, 10, 151-157, 307-311, 334-335
failure modes, 335
project prioritization and selection roadmap, 157-160
trends in, 164-165
strategic alignment, 342
time dedicated to, 230
Top-Down projects, 161-165
tracking. *See* software (for tracking metrics)
providers
accountability of, 58
contracts with, 67-68
differentiating, 63-66
identifying, 60-61
importance of, 55-56
partnership with, 59-60
questions to ask, 66-67
ROI (return on investment) of, 57, 59
selecting, 61-63

Q–R

QI Macros (desktop management software), 300
quality dimensions, reinvigorating Six Sigma, 333-334
quantifiable results of Six Sigma, 12

Ramelli, Donnee, 44, 110
recognition and reward programs, 274-279, 323
reducing head count, 92
reengineering, Six Sigma versus, 57
The Reengineering Revolution (Hammer), 57
Reengineering the Corporation (Hammer), 57
references, differentiating Six Sigma providers, 64
Request for Proposal. *See* RFP
requirements for enterprise management software solutions, 290-297
resource management, 294, 317, 321. *See also* infrastructure
results
quantifiable results of Six Sigma, 12
tracking, 295-296
retention of Belts, 279-281
return on investment. *See* ROI
reviewing
projects, 179-182, 321-322
Six Sigma assessment example, 326-328
reward and recognition programs, 274-279, 323
rework (metric), 99
RFP (Request for Proposal)
evaluating providers, 60
sample of, 357-365
RFP decision matrix, selecting providers, 61-63
roadmaps
delivering money, 35-36
leadership roadmap, 150, 303
management support, 319, 321-323, 337-338
process improvement roadmaps, 318-319, 336-337
project selection, 151-157, 307-311, 334-335
steps in, 306-307
sustaining performance, 323-325, 339-340
training plans, 312-317, 335-336
project prioritization and selection roadmap, 157-160

ROI (return on investment)
factors affecting, 75-80
of Six Sigma providers, 57, 59
Rosansky, Victor, 70, 309
RTY (rolled throughput yield), 87, 93

S

sales department, extending Six Sigma to, 348
Sauer, Brad, 46, 126
scheduling Six Sigma deployment events, 113-114
Schroeder, Richard, 26, 45, 118, 126
scope of deployment, 105-106
business unit versus division deployment, 109-110
connecting Six Sigma with other initiatives, 106-108
domestic versus global deployment, 110-111
integration of Six Sigma programs, 112-113
integration with Lean Enterprise, 111-112
pilot versus full deployment, 108-109
scheduling events, 113-114
scrap (metric), 97-98
selecting
Initiative Champions, 45-46
projects, 10, 151-157, 307-311, 334-335
failure modes, 335
project prioritization and selection roadmap, 157-160
trends in, 164-165
providers
differentiating providers, 63-66
questions to ask, 66-67
RFP decision matrix, 61-63
sense of urgency (deployment architecture stage), 42-45
Seven Habits of Highly Effective People (Covey), 56
short-term gains (deployment architecture stage), 51
Sierk, Jim, 6, 45, 126
SigmaFlow (desktop management software), 300
simulation applications, 299
Six Sigma
for alignment, 6-7
assessment example, 326-328
as change initiative, 6
connecting with other initiatives, 106-108
deployment. *See* deployment
for discipline, 7-8
forms of, 39
handbook example, 325-326
history of, 25-26
infrastructure. *See* infrastructure
integration of programs, 112-113
integration with Lean Enterprise, 213-215
as leadership development, 8-9
leadership roadmap. *See* leadership roadmap
methodology of, 10-12
organizational culture and, 37
process improvement in, 19-23, 73-75
providers. *See* providers
reengineering versus, 57
reinvigorating, 329-331
deployment efficiency, 340-341
extending Six Sigma, 346-352
leadership roadmap, 334-340
program assessment, 331-333
quality dimensions, 333-334
strategic alignment, 341-346
TQM (Total Quality Management) versus, 26-27, 56-57
training portfolios, 211-213
Six Sigma for Operations, 112-113
software (for tracking metrics), 285-287
desktop management solutions, 298-300
enterprise management solutions, 287-290
packages for, 297-298
requirements, 290-297
executing, 300-301

importance of, 344-346
planning for, 300
tuning and expanding, 301-302
Solso, Tim, 46, 126
sponsors. *See* Initiative Champions
statistical analysis software, 299
Steele, Richard, 69
steering teams, 46-47, 59, 142-143
strategic alignment, reinvigorating Six Sigma deployment, 341-346
strategic objectives, alignment to, 291
strategic planning, 69-85
strategy and vision (deployment architecture stage), 47-48
student evaluations of training plans, 230-234
success factors, Six Sigma deployment, 351
succession planning, 271-272
supply chain, extending Six Sigma to, 350-351
survey for Six Sigma assessment, 332-333

T

team members. *See* project team members
teams, resource management, 294
timing for deploymentt, 40-41
Top-Down prioritization matrix, 166-170
top-down projects, 10, 161-165
top-line expectations, 99-100
TQL (Total Quality Leadership), 107
TQM (Total Quality Management), Six Sigma versus, 26-27, 56-57
TQS (Total Quality for Speed), 107
tracking. *See* project tracking system
training. *See also* workshops
Black Belts (BBs), 133, 135, 225, 262-263
Business Team Workshop, 190-195
certification requirements, 144-146
Champion Workshops, 195-199
customizing workshops, 199-204
Deployment Champions, 127
Design for Six Sigma training programs, 215-218
Executive Team, 123
Executive Team Workshop, 186-190
Financial Workshop, 206
Green Belts (GBs), 136-137, 225, 262-263
Human Resources personnel, 141
Human Resources Workshop, 207
importance of, 119, 209-210
Initiative Champions, 125-126
integration of Six Sigma and Lean Enterprise, 213-215
K-Sigma Black Belt training program, 220-225
Master Black Belts (MBBs), 130, 132-133, 225-227, 262-263
materials review process, 204
Operational Black Belt training program, 218-220
portfolios for, 211-213
Project Champions, 129
project team members, 138
for sustaining performance, 324
training materials, 58
training plans, 312-317, 335-336
extending Six Sigma with, 350
failure modes, 336
management of, 227-228
number of Belts to train, 228-230
student evaluations, 230-234
time dedicated to, 230
training steering team, 130
Transactional Six Sigma, 39, 113
transfer of technology, 59
tuning software solutions, 301-302
"Turning Great Strategy into Great Performance" (Mankins and Steele), 69

U–V

validation of financials, 294
value mapping, 28-30
Visio (process map software), 299
vision and strategy (deployment architecture stage), 47-48
vision communication (deployment architecture stage), 48-49, 240-241

W–Z

waves of training, 262
Weidman, David, 6, 46, 122, 126
Welch, Jack, 26, 53, 57, 108-109, 122, 136, 173, 228, 245, 272, 285
Winning (Welch), 57
Winning Together program example (communications plans), 238-240
workshops. *See also* training
- Belt retention, 281
- Belt selection process, 266
- Business Team Workshop, 190-195
- Champion Workshops, 195-199
- customizing, 199-204
- Executive Team Workshop, 186-190
- Financial Workshop, 206
- Human Resources Workshop, 207
- materials review process, 204
- for Six Sigma/Lean Enterprise integration, 214-215
- tips for success, 205-206

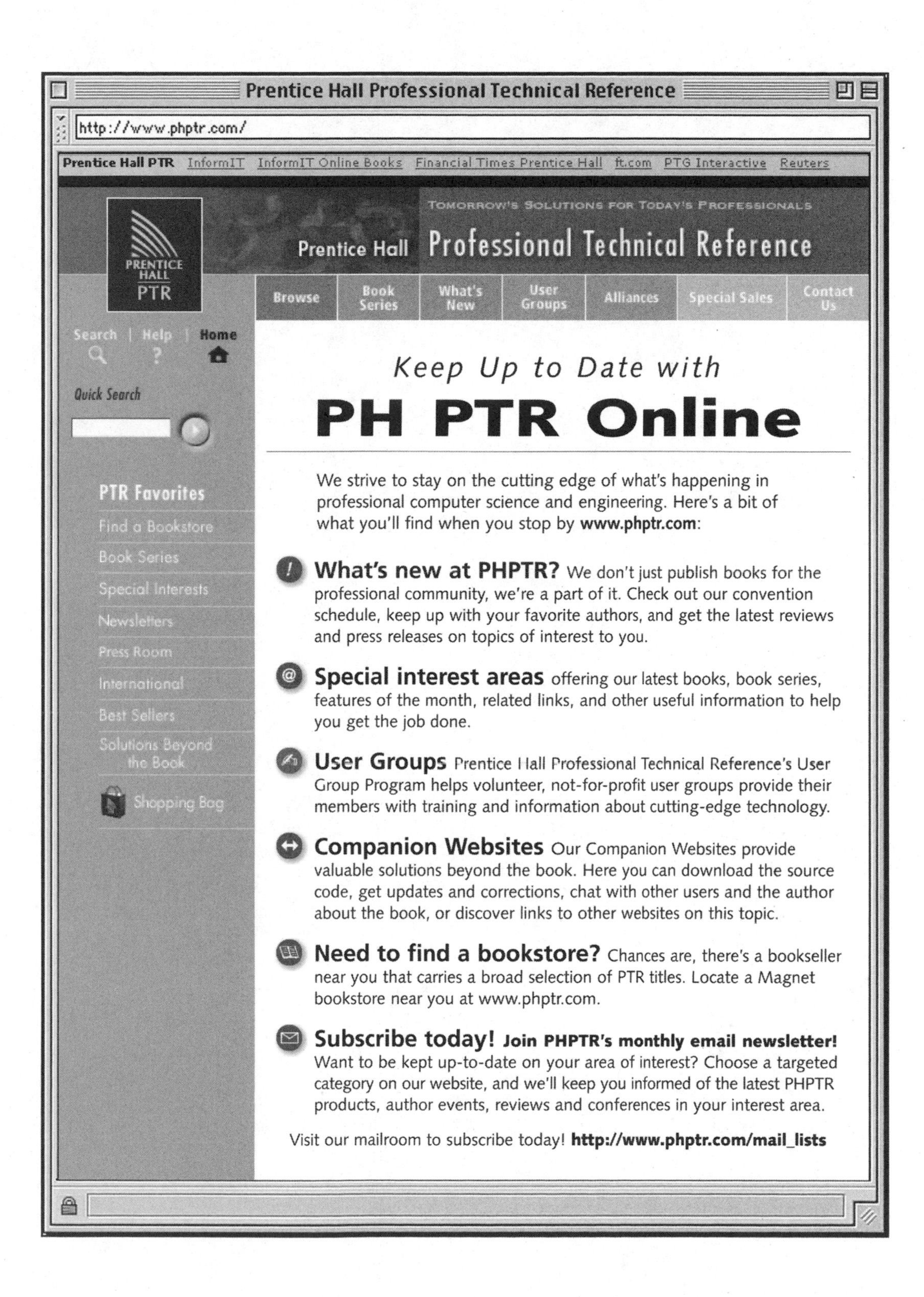

Prentice Hall Professional Technical Reference
http://www.phptr.com/
Prentice Hall PTR
InformIT
InformIT Online Books
Financial Times Prentice Hall
ft.com
PTG Interactive
Reuters
Tomorrow's Solutions for Today's Professionals
Prentice Hall
Professional Technical Reference
PRENTICE HALL
PTR
Browse
Book Series
What's New
User Groups
Alliances
Special Sales
Contact Us
Search
Help
Home
Quick Search
PTR Favorites
Find a Bookstore
Book Series
Special Interests
Newsletters
Press Room
International
Best Sellers
Solutions Beyond the Book
Shopping Bag
Keep Up to Date with
PH PTR Online
We strive to stay on the cutting edge of what's happening in professional computer science and engineering. Here's a bit of what you'll find when you stop by www.phptr.com:
What's new at PHPTR? We don't just publish books for the professional community, we're a part of it. Check out our convention schedule, keep up with your favorite authors, and get the latest reviews and press releases on topics of interest to you.
Special interest areas offering our latest books, book series, features of the month, related links, and other useful information to help you get the job done.
User Groups Prentice Hall Professional Technical Reference's User Group Program helps volunteer, not-for-profit user groups provide their members with training and information about cutting-edge technology.
Companion Websites Our Companion Websites provide valuable solutions beyond the book. Here you can download the source code, get updates and corrections, chat with other users and the author about the book, or discover links to other websites on this topic.
Need to find a bookstore? Chances are, there's a bookseller near you that carries a broad selection of PTR titles. Locate a Magnet bookstore near you at www.phptr.com.
Subscribe today! Join PHPTR's monthly email newsletter!
Want to be kept up-to-date on your area of interest? Choose a targeted category on our website, and we'll keep you informed of the latest PHPTR products, author events, reviews and conferences in your interest area.
Visit our mailroom to subscribe today! http://www.phptr.com/mail_lists